Strategic Decision Making in Modern Manufacturing

Strategic Decision Making in Modern Manufacturing

Harinder Singh Jagdev
*University of Manchester
Institute of Science and Technology (UMIST)
United Kingdom*

Attracta Brennan
*National University of Ireland
Republic of Ireland*

Jimmie Browne
*National University of Ireland
Republic of Ireland*

KLUWER ACADEMIC PUBLISHERS
BOSTON / DORDRECHT / LONDON

Distributors for North, Central and South America:
Kluwer Academic Publishers
101 Philip Drive
Assinippi Park
Norwell, Massachusetts 02061 USA
Telephone (781) 871-6600
Fax (781) 681-9045
E-Mail <kluwer@wkap.com>

Distributors for all other countries:
Kluwer Academic Publishers Group
Post Office Box 322
3300 AH Dordrecht, THE NETHERLANDS
Telephone 31 78 6576 000
Fax 31 78 6576 254
E-Mail <services@wkap.nl>

 Electronic Services <http://www.wkap.nl>

Library of Congress Cataloging-in-Publication Data

A C.I.P. Catalogue record for this book is available from the Library of Congress.

Strategic Decision Making in Modern Manufacturing
Edited by Harinder Singh Jagdev, Attracta Brennan, and Jimmie Browne
ISBN 1-4020-7497-2

Printed on acid-free paper.
Printed in United Kingdom by Biddles/IBT Global

Cover art designed by Attracta Brennan.

To:

- In the loving memory of my parents Balbir Singh and Prakash Kaur
  – I am sure we'll meet again
- Hardeep, Harjit and EG (my *bobykins*)
  – The gleams of sunshine of my life

- My parents Matt & Mona and Des, Elena, Bernie and David
  – You have left footprints on my heart

- Maeve, Lorcan, Shane, Ronan and Fergus
  – They give meaning and perspective to my life

Jab Tun Aaya Jagat Mein, Log Hanse Tu Roye
Aise Karni Na Kari, Pache Hanse Sab Koye
– Kabir

Contradictions do not exist. Whenever you think that you are facing a contradiction, check your premises. You will find that one of them is wrong.

– Ayn Rand *(Atlas Shrugged)*

Contents

Preface

The rapid pace of technological innovation and the effects of the Information and Communications Technology (ICT) revolution have resulted in dramatic changes on a global scale, from the empowerment of the individual to the spawning of global markets. From the business perspective, the widespread deployment of Information Technology (IT) has resulted in many organisational changes and the development and use of new management and business processes. An important challenge for today's manufacturing organisations is to be able to anticipate the impact of investments in new (frequently IT based) manufacturing technologies and programmes. Ideally, management need to be able to identify and articulate the many ways in which investment decisions influence their organisation - in terms of human resources and skills needs, organisational structures, resource requirements and performance across a range of measures. The underlying question therefore is; how can management both anticipate the effects, and measure the success of technological and/or programme investment decisions?

One answer to this question is to develop and use tools, which bridge the gap between strategic management considerations and the operational effects of technology investments on the manufacturing organisation, so that the likely impact of new manufacturing technology and/or programme implementations can be evaluated, anticipated and accurately predicted. Such bridging is important, because in today's manufacturing environment, it is increasingly necessary that a close relationship exists between manufacturing decision making and corporate business strategy, so that manufacturing decisions complement and are fully aligned with the organisation's strategic objectives.

The **AMBIT** (Advanced Manufacturing Business ImplemenTation) approach has been specifically developed to bridge this gap between strategic management and manufacturing management and, act as an enabler of manufacturing technology and programme investment decisions. AMBIT focuses specifically on the non-financial aspects of such investments and offers an approach which allows a manager, or more frequently a management team, to understand the impacts of a new technology or a new programme on the manufacturing organisation in terms of manufacturing performance.

The project which financed much of the development of the AMBIT approach was a fundamental research project funded by the European Union through its BRITE-EURAM research programme.

This book is structured in a way in which we believe shows the logical development of the AMBIT approach.

- **Chapter 1** presents an overview of the IT revolution and its general impact on organisational activity. It describes the importance of innovation, flexibility, strategy and managerial decision making. It reconsiders the traditional managerial role in the context of new managerial mindsets and competencies.

- **Chapter 2** summarises the key concepts relating to performance measurement. The underlying assumption of this chapter is that appropriate performance measurement is critical to success in any organisation because the use of inappropriate performance measures may lead to dysfunctional behaviour and consequently, poor organisational performance. This chapter compares traditional and contemporary performance measurement systems and introduces the reader to currently available performance measurement approaches.

- **Chapter 3** presents an overview of the AMBIT approach. It describes the importance of articulating a clear manufacturing strategy in the development of a competitive position. Furthermore, it outlines and discusses the reasons for developing the AMBIT approach.

- **Chapters 4** through **6** inclusive, detail the structure and the three key stages of the AMBIT approach.

  o   In chapter 4, the AMBIT performance measurement framework and the MOMP (Measures of Manufacturing Performance) map are presented. We believe that these represent some of the more critical features of the AMBIT approach. The AMBIT performance measurement framework shows how the organisation's business goals (expressed in terms of its critical success factors (CSFs)) can be linked to its manufacturing strategy by means of the measures of manufacturing performance (MOMPs).

  o   In chapter 5, the manner in which the technology and programme issues are linked to the manufacturing strategy is described. We also outline the technologies and/or programmes which management might wish to implement in the manufacturing plant. We do this by using an example technology, namely *interconnect technology* (used in the electronics assembly industry) and an example programme, namely *concurrent engineering*. A proposed approach for the linkage of the technology and programme issues to the manufacturing strategy (and performance) is outlined. Such a linkage is necessary in order to assess the impact of the proposed technology an/or programme on manufacturing performance.

  o   In chapter 6, a selection of analysis tools which can be used to support the manager or team of managers in evaluating the strategic and operational effects of a particular technology or programme implementation are introduced. These tools include Analytic Hierarchy Process (AHP), Case Based Reasoning (CBR) and Semantic Modelling.

- Finally, Chapter 7 introduces the reader to ideas and concepts that might prove useful in developing an effective computerised tool-set to support the AMBIT approach. In this chapter, the manner in which individual learning and creativity are facilitated through the design of the tool is examined.

As the goal of the AMBIT approach is to act as an enabler of manufacturing technology and/or programme investment decisions, we have used various example programmes and technologies to validate the overall approach. Whilst interconnect technology, concurrent engineering, business process re-engineering and lean production are used as examples of technologies and programmes to illustrate the workings of the AMBIT approach, it is important to stress that there is no particular significance attached to their choice. Programmes such as Total Quality Management (TQM), Lean Manufacturing and Flexible Manufacturing Systems (FMS) could just as easily have been selected.

We see AMBIT as an approach which allows a manager or a management team to consider and articulate in advance, the likely consequences of major technology and/or programme decisions on a manufacturing organisation. Clearly AMBIT, if it is to be truly effective, needs to be supported by appropriate software tools and (ideally) an integrated tool-set. The development of such at tool–set is beyond the scope of this book.

The prediction of future outcomes is a very imprecise and ambiguous activity at the best of times. Yet despite such ambiguity and uncertainty, it is no longer wise for managers to steer the organisation into the future by looking over their shoulder. Managers need to be forward looking. They also need appropriate tools and approaches to support them in their task. Thus, whilst the pages of organisational history may be filled with anecdotes about organisations which failed to anticipate the future, it is the challenge of today's organisations to evade such a fate.

H.S. Jagdev, A. Brennan, J. Browne
May 2003

Acknowledgements

Much of the work reported in this book was completed during a number of EU and industrially funded projects. In particular the EU Brite-Euram funded basic research project entitled AMBITE (Advanced Manufacturing Business and Information Tool for Europe, Project Number: BE 7094) was critical to the development of the ideas outlined in this book. We would like to acknowledge the contributions of our partners in AMBITE, in particular Peter Sackett and his team at Cranfield Institute of Technology and Rudiger Dillmann and his team at the University of Karlsruhe. Other colleagues with whom we have worked closely over recent years have also greatly influenced our thinking and will recognise some of the influences in the various chapters. We are thinking particularly of colleagues in various universities around the globe, including: Asbjørn Rolstadås and Bjørn Anderson (University of Trondheim), Bernd Hirsch and his team (University of Bremen), Guy Doumeingts, Jean-Paul Bourrières and Bruno Vallespir (University of Bordeaux), Hans Wortmann and his colleagues (Technical University of Eindhoven), Eero Eloranta (Helsinki University of Technology), Claudio Boer (Politecnico di Milano), Peter Falster (Technical University of Denmark), Eamonn Murphy (University of Limerick), Nicos Karcanias (City University, London), Petros Groumpos (University of Patras), Allan Carrie and Umit Bittici (University of Strathclyde), Stephen Childe (University of Plymouth), Norman Sadech (Carnegie-Mellon University), Collin Moody (Purdue University), Peter O'Grady and Andrew Kusiak (University of Iowa), Dan Shunk (Arizona State University), Roger Nagel (Lehigh University), Thomas Gulledge (George Mason University), Nallan C Suresh (State University of New York, Buffalo), Zongchi Chen (Beijing Institute of Aeronautics and Astronautics) and Chen Yuliu (Tsinghua University).

Several colleagues and former students of CIMRU in NUI, Galway (Paul Higgins & David O'Sullivan) and UMIST (Colin Walters & Xiao Jun Zeng) have influenced the development of ideas presented in this book. However, our particular thanks must go to Padraig Bradley, Tommy Dolan, Séan Jackson and Brian Kelly; whose work has had direct impact on the AMBIT approach.

About the Authors

Harinder Singh Jagdev graduated from Indian Institute of Technology, Madras, in Mechanical Engineering in 1974. After working for two years as Production Engineer at Mercedes Benz, he joined UMIST, where he gained M.Sc. and Ph.D. in Manufacturing Technology from Victoria University of Manchester. Since 1980 he has researched for and been consultant to many organisations in Europe. He is also very active in the EU-funded research programmes, both as a proposal and project reviewer on behalf of the Commission and member of the research consortia. He is the editor of journal Computers in Industry and also serves the editorial board of Production Planning and Control.

Attracta Brennan is a lecturer in the department of Information Technology at the National University of Ireland, Galway. She holds a Ph. D. degree from CIMRU, at NUI Galway. Her research interests are in the areas of design, creativity, learning and innovation management.

Jim Browne is Registrar and Deputy President of the National University of Ireland, Galway, as well as the founder of CIMRU there. He is also a former Dean of Engineering in Galway. He holds Bachelor and Master's degrees from NUI Galway and Ph.D. and D.Sc. degrees in engineering from UMIST in Manchester. He has authored or co-authored more than 200 papers, and co-authored or co-edited 11 books. His research interests are in the modelling, design and analysis of extended enterprises and virtual enterprises.

Chapter 1

Strategic Management in the Global Marketplace

Keywords: Societal effects of IT (information technology), IT overview, technology, global marketplace, product life cycle, management, integrative cultures, segmentalist approaches, organisational strategy, strategic planning, the strategic thinker, managerial decision making, decision categories, decision models, strategy and information technology, knowledge management.

Chapter Objectives: With the explosion of available information technology based products, the IT revolution has resulted in information accessible at our fingertips. The widespread access to this global information has resulted in increased competition and intensified the challenge for today's organisations to survive and prosper. Complacency in the marketplace is no longer acceptable. Competition has intensified, whilst the marketplace itself has become more discerning, varied and dispersed. Indeed, the very nature of the manufacturing climate has inspired new approaches with respect to the organisational structure, the managerial role and manufacturing procedures. Many believe that long-term organisational success is greatly dependent on a flexible strategy. However, the challenge to today's organisation in achieving its strategic objectives rests to a large extent on IT and the proper management and presentation of information, such that organisational decision makers are facilitated in distinguishing between realistic alternatives and mis-information. Clearly, managers require IT based tools to support and facilitate them in what is considered one of the most critical of managerial roles - strategic decision making. Having read this chapter, the reader should become familiar with:

- The global marketplace and its characteristics, some of which include:
  - Increased product diversity.
  - Reduced product life cycles.
  - Increased focus on business process re-engineering (BPR).

- The practice and importance of management and the forces, which are making the task of management more difficult
- The difference between the segmentalist management style and the integrative management style, and the relevance of the latter for today's organisation.
- The different levels of strategy (i.e. corporate, business and functional strategy) and the whole process of strategic planning.
- The influence of information technology (IT) on an organisation from a strategic viewpoint.
- The difference between knowledge, information and data.
- Why managers use information systems and what they really require from such systems.

1.1   An Overview of Information Technology (IT)

Because information technology (IT) promotes communication on a global scale, it is enabling a force which is shaping and will continue to shape mankind's future i.e. the social power of people. The desires of people and their expectations will, to. a large extent, fashion the future. Thus, the very fact that IT promotes increased individual communication on a global scale shows that IT is a major driver of change. From an organisational viewpoint, IT is critical for the effective and efficient response to changing customer needs and competitive pressures. IT can also lead to enhanced relations between organisations and their customers through the sharing of information to geographically remote customers. It supports management in their managerial role and enhances the capacity to innovate. IT has also proven beneficial in the task of process flow management. Clearly, the application and usage of IT based tools and technologies has strong implications for any organisation. "It is frequently argued that IT is the most important factor in increasing productivity and reducing costs" [Dewhurst *et al.* 2003].

Since man first appeared, he[1] has made tools and weapons of one kind or another to support him in his fight for survival. Tools enable him to perform tasks without whose use he would be incapable of performing. Tools also assist him in executing a task better or faster than he would be able to, prior to their use. This perpetuation of tool development represents one way in which man is distinguishable from the other animals. With the passage of time, the relationship between man's tools and his knowledge has grown into what is called his technology. Technology is basically that which assists man in controlling and creating the circumstances in which he operates. It can be supportive or it can be prohibitive. It can be the driving force of increasing employment and competitive ability, yet it can also represent the source of major economic disruption. Within the organisational context, the objective of IT is to ensure that the information required by an organisation to achieve its goals, is made available. IT does so by enabling the collecting, processing, storing and distribution of organisational data.

---

[1] Please note that no distinction to gender is made throughout this book. Therefore, the terms such as man/woman, he/she, him/her, his/her, etc. are interchangeable.

1.2   Organisations and the Global Marketplace

Information as a worldwide commodity has greatly increased our understanding of the world around us and brought a high standard of living and economic prosperity. It has also challenged the basis on which organisations operate. Table 1.1 highlights the developments in the competitive market from the 1960s through the 2000s and the corresponding developments in market requirements and product strategy.

Table 1.1 The link between process innovation and market developments [after Brown *et al.* 1996]

Decade	Market Characteristics	Market Requirements	Product Strategy
1960s	Stable	Price	Few varieties
1980s	New competition	Price, Quality	Quality
1990s	Increased competition	Price, Quality	Differentiation, Quality, Wide range
2000s	Volatile, Dynamic	Price, Quality, Uniqueness	Constant flow of new products, Differentiation

This table suggests that many markets in the 2000s (due to economic conditions and consumer demands) are characterised by their volatility, rapid change, reduced product lifecycles and increased speed of new product introduction. This represents a fundamental shift from a 'producer offered' to a 'customer driven' approach. The adoption of a customer driven approach is vital if organisations are to stand any hope of competing successfully in today's markets [Chan *et al.* 2003].

From the manufacturing organisation viewpoint, it is important that all efforts, processes and infrastructure are geared to satisfying customers' needs. This means that processes have to be flexible and capable of producing high variety (Figure 1.1). Such an approach must be central to the organisation's manufacturing strategy.

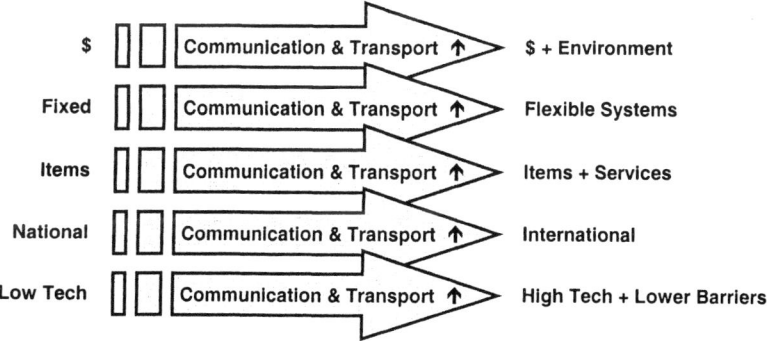

Figure 1.1 The new environment for manufacturing [after Pfeifer *et al.* 1994]

With the need to satisfy customer demands and remain competitive, organisations need to (Figure 1.1):

- Include (as a result of increasing changing values ↑) environmental considerations as part of their competitive arsenal.
- Compete globally as opposed to just locally.
- Adopt more flexible systems because of the increased rate of technological progress.

Organisations increasingly operate within a manufacturing climate, distinguished by:

- Increased product diversity[2].
- Reduced product life cycles.
- Increased focus on business process re-engineering (BPR).

1.2.1 Increased Product Diversity

> Choices, choices, choices. We all have them. Our competitors have them. More importantly, our customers have them!   – Horowitz *et al.* 1995

Because today's consumers are more discerning and demanding, they no longer just accept whatever is given to them. Such discernment has propelled manufacturing into the era of the personalised consumer product. Today's organisations are as a result, faced with the challenge of providing a wide variety of products at reduced prices, whilst still remaining competitive which can result in "whole markets appearing, disappearing and mutating at an alarming rate" [McCarthy 2003].

This increase in product variants has resulted in the following:

- **Demand fluctuations**. The forecasting of production and materials requirements becomes more complex with the growth in product variants and the inability to anticipate the mix of orders which will come in.
- **Capacity complexity.** An explosion in the number of product variants can result in increased production complexity with smaller batches, more set-ups and more changeovers required.
- **Collaborative Design.** With organisations espousing the extended enterprise model, they are increasingly investing in collaborative design projects. Accrued benefits can include; reduced design process and design errors, faster time to market and enhanced communication with suppliers. Collaborative design requires the deployment of IT tools and technologies to support both design information and design process sharing [Dewhurst *et al.* 2003].
- **Data explosion.** Because the specifications of each product variant have to be described in detail, the volume of information mushrooms.
- **Inventory factors.** As product variants generally require different components, more inventory items may need to be managed.

---

[2] IT enhances an organisation's ability to innovate and generate product variants through the use of IT based tools/technologies (such as CAD) and the speedy transmission of information amongst the members of a design team [Dewhurst *et al*, 2003]. The latter is increasingly important given the dispersed nature of the design team (either departmentally or geographically) in the virtual enterprise (see section 3.5).

- **Resource issues.** With product variants competing for similar production resources (i.e. people, machines and materials) scheduling and sequencing gain added complexity.

> Henry Ford would be an utter failure telling today's customers that they can have his cars in any colour they want, as long as it's black.   – Fraser 1995

1.2.2 Reduced Product Life Cycles

The product life cycle is a means of tabulating the rise in popularity and the eventual 'death' of a product. The life cycle phases include: product introduction, product growth, product maturity and product decline (Figure 1.2). Whilst this model (which has its inception in the 1950s) appears to some extent, a simplistic approach to the representation of the product life cycle, it still provides a clear illustration of product evolution [Reddy *et al.* 2002].

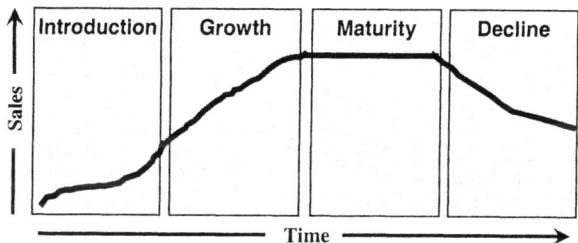

Figure 1.2 Product life cycle

- **Product introduction.** The product is introduced and launched onto the marketplace. A high acceptance of the product will generally result in a brief introductory stage. Brown argues that some of the reasons which prompt new product introduction include [Brown 1996]:
  o Satisfying customer demands.
  o Because of the intensely competitive nature of the manufacturing environment, organisations are compelled to continuously generate and launch new products (e.g. electronics consumer products).
  o Sustaining present sales levels. These levels may not be possible with current products when they reach the end of their product life cycle.
  o The organisation might enjoy a first to market reputation, which it wants to continue.
  o The improvement of organisational profitability.
- **Product growth** represents an acceptance of and demand for the product. It is characterised as a period of high sales, resulting in increasing profits, which all too soon become shadowed by increasing competition.
- **Product maturity.** As the challenge of competition increases, sales level off. In an attempt to extend the product's life, product variations and modifications are introduced into the marketplace with greater frequency.
- **Product decline** occurs with falling sales and reduced product demand. Decline represents the end of the product life cycle.

Traditionally, organisations would be able to anticipate many years of stable demand for their products. During such periods, organisations could recover the costs incurred during the product and process development stages, as depicted in Figure 1.3.

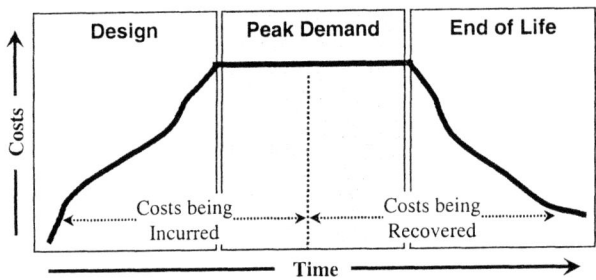

Figure 1.3 Distribution of cost over the product life cycle

However, such an expectation of high, stable demand is not always practical today. Demand is variable. Products are continually being re-designed and their life expectancy is continually being challenged by market competitors. In an attempt to counteract such market instability, organisations are compressing the time taken for product design and product manufacture, in order to get the product promptly to the market. This haste has called for flexible production systems to help cope with existing product design and future product re-design. However, whilst organisations need flexible production systems, they also need cost-effective production systems. Such systems are required so that the rewards accruing during the much shorter peak demand stage of the product life cycle will be much greater than the costs incurred at the product design and process development stages (Figure 1.3).

> Manufacturing has had to maintain improvements in productivity and cost, but also monitor (sense) emerging demands, and when required, rapidly re-configure an appropriate organisational system (respond).   – McCarthy 2003

1.2.3   Increased Focus on Business Process Re-engineering

The focus on customised products represents the driving force of current manufacturing organisations. It emphasises the fact that consumers, rather than suppliers, have become the governing force in the *supplier-consumer chain*. Whilst this has resulted in high quality and customisable products produced at low prices, it has also meant that if organisations want to remain competitive, they must be able to compete on the basis of product variety, quality and price. Therefore, the challenge for organisations is to be able to adapt to new consumer buying patterns and respond to aggressive competition. To do so, organisations are re-engineering their businesses to create flexible, responsive and customer-focused business processes.

1.3 Management and Strategy

> The move towards the information society presents a competitive challenge to which all parts of industry need to respond; to adapt to new ways of thinking and working; to come up with new products and services; to adapt to new structural relationships with suppliers and customers and to find additional customers in new markets. – Lang 1996

In response to the increasing fluidity of global markets, which are requiring organisations to take action against the "challenges from new competitors, partners, new products and new technologies at an ever increasing pace" [Reddy *et al.* 2002], the effective management of today's organisation will involve a variety of skills, one of which is an emphasis on creative management. Increasingly, creative management is emerging as a growth area with its support of "greater behavioural competence in taking the initiative and responsibility under conditions of risk and uncertainty, greater perceptual competence in gathering and organising information and taking the perspective of different organisational sub-units, greater affective competence in empathising with others and in resolving conflicts among managers with different viewpoints and greater symbolic competence in one's ability to conceptualise the organisation as a system are all required to make modern organisational forms work effectively" [Kolb *et al.* 1986].

Whilst increased innovation and technological adoption have resulted in shorter product life cycles and high requirements to be met by both products and services, the changing nature of time within the organisational context has signified that strategic decision making is seriously impacting organisational prosperity. Managers have to make effective strategic decisions within ever collapsing time frames.

1.3.1 The Importance of Management

> The pivotal event which shaped the world into the form we now see around us was the British Industrial revolution, which began in the late eighteenth century. Before it, most societies were based on small-scale, self-sufficient agricultural production, with the vast majority of the population, some 80 to 90 percent, living in the countryside. By the end of the nineteenth century, after the Industrial revolution had run its course, the reverse became the case, in the leading industrialised countries at least, with most people living in urban centres and depending on industrial and commercial activities for their livelihood. – Burnes 1996

Even though the practice of management can be traced back to 3000 BC, it was not given serious attention until the emergence of large organisations in the late 1800s. The art of management practice involves planning, decision making and controlling an organisation's resources (i.e. human, financial and informational). As a result, management really refers to the administration and guidance of an organisation such that the organisation is as successful as possible. Successful management is critical to organisational prosperity and longevity. However, the task of management is made more complex with the influence of economic, social, environmental and political forces:

- **Economic forces**. Product variability, technological variance and the competitive global market place intensify the pressure of the management task.

- **Social forces**. It is widely accepted that in the early years of mass production and industrialisation, many employees were poorly treated. However, with the IT revolution and the resulting empowerment of the individual, social forces have developed to influence the emergence of a different style of working and of organisation. Such developments have also affected management thinking in areas such as motivation and leadership. If an organisation fails to support its employees in the development of their individual skills and qualities, it will be unlikely to achieve its full potential.
- **Environmental forces**. With the growth in public concern over environmental issues and the proliferation of green consumerism, sensitivity to environmental factors is becoming increasingly important for today's organisations. As a result, for very many industry sectors, it is important that the current generation of managers be influenced by and educated in environmental issues. "Environmentalism is not a new political cause; its roots reach deep into the nineteenth century and counter-reactions to the Industrial Revolution. Prior to the industrial revolution, pollution was barely recognised as a political issue. Enloe notes that in the seventeenth century, the Stuarts were petitioned about the environmental dangers to Londoners arising from coal burning in the capital, although legislative action was not taken until the enactment of the Clean Air Act almost two centuries later. In effect, it was not until the late nineteenth century and early twentieth century that environmental issues became politically salient". [McGrew 1993]
- **Political forces**. The introduction and intensification of legislation with respect to environmental issues, health and safety, employee rights etc., is significantly impacting management.

The influence of these external forces is illustrated in Table 1.2 (which represents an historical review of the product market). This Table suggests that the marketplace of the 1950s was characterised by moderate competition, high demand and mass production, with suppliers as the driving force in the supplier-consumer chain. However, increased competition, globalisation, employee consideration and environmental awareness are characteristics of the marketplace of the 1990s and the post 2000s.

Harris and Browne *et al.* suggest that the following changing market forces strongly influence manufacturing systems [Harris 2002, Browne *et al.* 1994]:
- **Globalisation**. The internationalisation or globalisation (which has been described as the "movement of business, industrial and professional activities into a global marketplace" [Harris 2002]) of markets has resulted from the reduction in trade barriers, the opening of new markets, the advances in information technology and the improvements in mass transportation. Competition on a global platform affords organisations from geographically diverse areas of the world, the opportunity to compete in any marketplace. As a result, today's market is experiencing decreasing product life cycles and improved services, with respect to enhanced product information, training and repair services etc. In response to this competitive climate, organisations are becoming more flexible and customer oriented, and are moving away from a hierarchical structure to a structure dominated by teams.

- **Environmentally benign production**. Societal awareness of the need for environmentally benign production is resulting in pressure on manufacturers to create production systems which accommodate end of life disassembly and resource recovery.

Table 1.2 Historical review of the product market [after Pfeier *et al.* 1994]

1950s	1970s	1990s	2000s
High demand	Strong competition	Internationalisation	Rapidly changing markets
Moderate competition	Increasing requirements	Employee consideration	Legal restrictions
⬇	⬇	⬇	⬇
Mass production	Buyers market	Shorter innovation periods	Market specific variants
Broad market	Increasing variants and complexity	Environment / Recycling friendly products	Globalisation

Classical management practice used the Taylorist approach in the improvement of individual worker performance. This approach held the view that employees deliberately worked within their full potential and were solely motivated by money. Employees worked to advance their own interests and intentionally kept employers oblivious to the speed at which work could be carried out. They would frequently try to achieve the maximum reward for the minimum work (referred to as 'soldiering' by Taylor). They would use whatever bargaining power their skills or knowledge would allow to ensure this end. In addressing this situation and optimising employee productivity, Taylor proposed the division of labour and the explicit definition of the steps to be taken for each stage of the task.

As Burnes [1996] points out "The workers become 'human machines', told what to do, when to do it and how long to take. More than this, however, it (the Taylorist approach) allows new types of work organisation to be developed and new work processes and equipment to be introduced; thus workers move from having a monopoly of knowledge and control over their work to a position where the knowledge they have of the work process is minimal and their control is vastly reduced. The result is not only a reduction in the skills required and the wages paid, but also the creation of jobs which are so narrow and tightly specified that the period needed to train someone to do them is greatly reduced. This removes the last bargaining counter of labour: scarcity of skills".

The Taylorist view of the human as a cog of the organisational machine was challenged by the growth of the Human Relations movement. This movement stemmed from the Hawthorne studies, which were carried out by Elton Mayo, and concerned the correlation between the level of lighting and worker productivity.

These studies placed more emphasis on people than machines and worked primarily on the principle that:

- Employees are emotional rather than economic-rational beings. As a result, the emotional and social needs of the employees had a greater influence on their work performance than financial reward schemes.
- Organisations are social rather than mechanical systems. Some of the emotional and social needs of employees are met through the formation of informal but significant social groups.
- In addition to formal practices and procedures, organisations comprise informal structures and rules. Employees create and use informal structures and patterns of behaviour in order to meet their own emotional needs. The satisfaction of these needs can have an influence on individual performance and can impact the organisational performance to a greater degree than formal structures and control mechanisms laid down by management.

A manager's concern for his employees will and does result in increased employee satisfaction and improved employee performance. Thus, whilst a manager is someone who plans and makes decisions, organises, leads and controls human, financial and information resources, the successful manager also views his employees as important organisational investments.

As suggested earlier, today's organisations greatly require management competencies which deal with rapid change. Organisations which are capable of successfully responding to change reflect what are called integrative cultures and structures. In an integrative organisation, problems are treated as wholes and the ramifications of actions are considered from an holistic viewpoint. Integrative management encourages reduced conflict and the exchange of information and new ideas (supported by IT tools and technologies) across organisational boundaries. This management style supports and nurtures creativity. In contrast, the segmentalist management style is anti-change oriented. A segmentalist approach focuses on the compartmentalisation of actions, incidents and problems. In this approach, problems, which are viewed as isolated events (non-related to any other problem) are solved by breaking them down into their component parts. These parts are then assigned to specialists working in isolation. As a result of such an isolated compartmentalisation approach, segmental organisations generally find it hard to be either creative or adaptable to change. Indeed, changes are often isolated in one segment and not allowed to touch other segments. This is done in order to reduce the disturbance that such change can bring. "In searching for the right compartment in which to isolate a problem, those operating segmentally are letting the past - the existing structure - dominate the future. The system is designed to protect against change, to protect against deviations from a predetermined central thrust and to ensure that individuals have sufficient awe and respect for this course to maintain their role in it without question - though they may fight over their share of the proceeds. As soon as a problem is identified, it is surrounded and isolated. Each person, each department, each level has only a part of any problem, and no assumed need to worry about any other part. Over specification of resources leaves little slack for experimentation" [Kanter 1995].

The change-oriented nature of today's organisation encourages integrative thinking. However, even though this supports organisational survival, today's management faces an uncertain future:

- A future where product, price, features and quality are no longer stable foundations for success.
- A future where organisations are consumer driven and consumers are increasingly discerning.
- A future where organisational employees may be dispersed geographically due to the nature of the organisational structure.
- A future where time is now the equal of cost and quality, with time to market, time to respond and time to implement being critical.
- A future where new technologies offer new opportunities.
- A future which requires that managers have a clear definition of the organisational strategy, such that the organisational activities and the employees work towards common goals.
- A future where management need increasing support in strategic decision making.

1.3.2   The Importance of Organisational Strategy

Get the facts first, you can always distort them later.   – Mark Twain

Whilst the view into the future has never been less explicit, the necessity of a cohesive organisational strategy, to maintain an economic and competitive advantage has never been greater. The three most popular forms of organisational strategy are Corporate, Business and Functional Strategy [Burnes 1996, Hayes *et al.* 1984], as depicted in Figure 1.4.

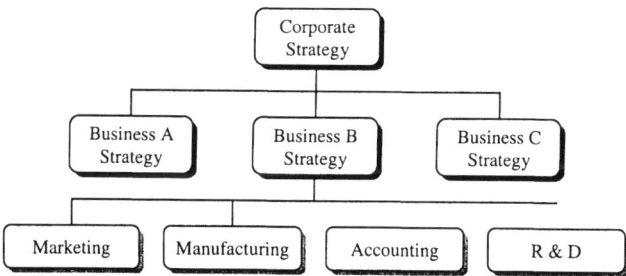

Figure 1.4 The three levels of strategy [after Hayes *et al.* 1984, Kelly 1995]

- **Corporate Strategy** distinguishes the business(es) in which the organisation participates, dependent on product, process and/or material diversity. The derivation of the corporate strategy comprises the following (Figure 1.5):
  1. Identify the organisational mission. The mission statement represents a statement of the organisation's purpose.
  2. Identify the organisational objectives. The objectives are identified by defining the goals of the different functional units (i.e. finance, manufacturing, product design and marketing etc) of the organisation.

3. Scan the environment. The environment is scanned by examining issues such as the economy, government policies, legal issues, competitor analysis and socio-environmental factors. Environmental scanning is done to heighten familiarisation of the environment in which the organisation functions and competes.
4. Evaluate the organisation's strengths and weaknesses.

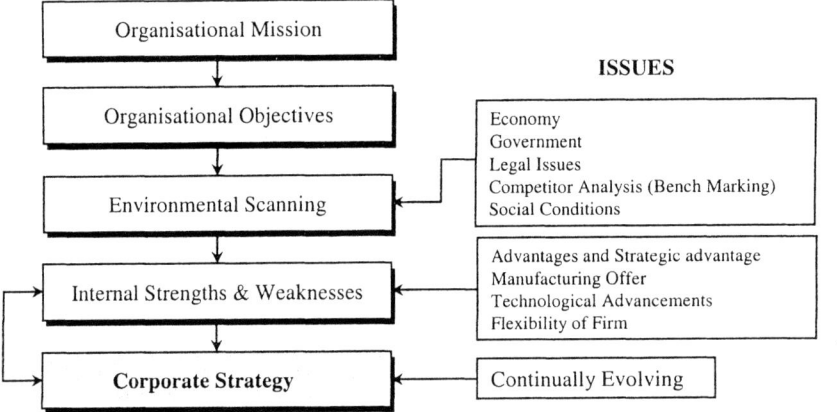

Figure 1.5 Formulation of corporate strategy [after Pannesi 1990]

The essence of strategy is winning. In the business arena, winning means creating superior customer value and dominating in the global marketplace.
                                                                        – Ma 2003

- **Business Strategy** specifies the scope and direction of each of the individual businesses and the basis on which these businesses achieve and maintain competitive advantage. The questions posed at this level include:
  1. How should the organisation position itself, such that it can successfully compete in distinct and strategically relevant markets?
  2. Which products should it provide to which set of customers?
  3. How should the internal aspects of the business be structured and managed, in order to support its specified competitive approach?

Business level strategies refer to the different ways that an individual business unit can compete in its chosen market(s). Business level strategies include:
1. Cost leadership. The objective of cost leadership is to have costs lower than one's competitors, without incurring a corresponding reduction in product quality.
2. Product differentiation. The aim of the product differentiation strategy is to provide customers with a diverse and superior range of products and services, in comparison to one's competitors.
3. Specialisation by focus. Specialisation by focus is concerned with the selection of particular markets, products or geographic areas in which to compete.

- **Functional Strategy** concerns individual business functions and processes, for example personnel and marketing. The individual functional strategy must be coherent with its respective business strategy. The questions posed at this level include:
    1. How can the strategies, which were formed at the corporate and business levels, be translated into tangible operational terms, so that the individual organisational functions and processes are supported in pursuing and achieving them?
    2. In what manner should the individual functions and processes of the business organise and structure themselves in order not only to attain their own objectives, but also to ensure that they integrate with the rest of the business in the creation of synergy?

Webb defines strategy as the "process of deciding a future course for a business and so organising and steering that business as to attempt to bring about that future course" [Webb 1989] (sic). The term strategy originates from the Greek "strategos", which is literally translated as General of the Army. The word general comes from stratos, an army and aegin, to lead. In ancient Greece, each of the tribes annually elected a strategos as commander of its regiment. At the battle of Marathon in 490 BC, it was the responsibility of the strategoi to provide strategic advice concerning the management of battles to win wars, as opposed to tactical advice on the management of troops to win battles. As the focus of the strategoi was holistic i.e. the big picture, the art of strategy was, and still remains, synonymous with the approach of grand design, plan, planning, policy, procedure, programme and scheme.

In more recent times, the Harvard Business School developed the Harvard Policy Model in the early 1920s. This model represented one of the first strategic planning techniques for private businesses and was mostly concerned with organisational policy and structure. However, during the 1950s, strategic planning concentrated more on risk management and market share and less on organisational policy and structure. Today, the development of an organisational strategy is a perpetuating process whereby the organisation and its environment are assessed, within the context of optimum exploitation of market opportunities. An organisational strategy is a dynamic and comprehensive plan, which is developed to support the achievement of the organisational mission. The importance of such a strategy is emphasised by Alkhafaji who concedes that "those who cannot adapt to new technology, changes in foreign policy and cultural differences will be dropped from the world market as others who plan and implement better strategies take their place. Lack of planning in the global market can mean loss of profits or even failure to survive" [Alkhafaji 1995]. This view is also supported by Stoney [2001], who agrees that strategic management is based on the assumption that the mutable nature of the global environment demands that the organisation be adaptable, flexible and responsive. If the organisation does not embrace these traits then it will not survive.

The activities comprising the strategic management task include:
1. Corporate strategy assessment. This involves an evaluation of the organisation's strengths and weaknesses, with potential opportunities and

threats also being identified. It represents a strengths, weaknesses, opportunities and threats (SWOT) analysis. The corporate strategy assessment is carried out in order to first evaluate the organisation's position relative to the external environment and then to select the most appropriate strategic alternatives with which to support the organisation within this environment.

2. Strategy formulation. Once realistic organisational goals are formed, a strategy/strategic plan is then identified which bests satisfies these goals.
3. Strategy implementation. This represents one of the most critical steps in strategy management. A strategic plan which is formulated correctly but incorrectly implemented will not produce the intended results. Therefore, it is important that the implementation strategy posses an operational plan which takes into consideration the organisation and movement of resources when they are required. Implementation of the strategic plan is also a means of validating the assumptions made during the planning process.

If an organisation existed within an environment characterised by an absence of change, then logically there would be no need for a strategic plan. However, this is not true of today's manufacturing environment, which is experiencing ongoing changes on a demographic, economic and cultural basis. Strategic planning is necessary. It represents a 'tool' which supports the organisational response and adaptation to change, so that an organisation can create a viable future for itself within such a context of change.

Strategic Planning

The future ain't what it used to be.   – Arthur C. Clarke

The confused organisation depicted in Figure 1.6 represents an organisation which does not have an overall strategic plan i.e. a strategic direction. The confused organisation is characterised by a lack of clarity with respect to the organisational objectives. This lack of overall focus often results in the individual business units attempting to improve the organisational performance by doing what they perceive to be most effective for their particular function, as opposed to planning for the overall organisation. Such disparate and often contradictory actions frequently result in disharmony and acrimony amongst directors and functional managers.

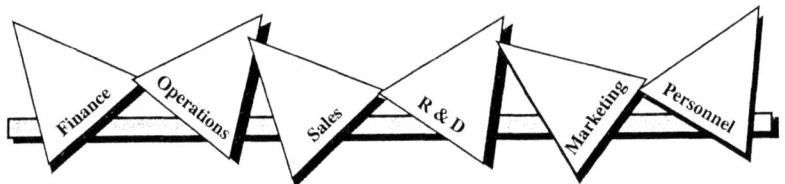

Figure 1.6 The confused organisation [after Reading 1993]

Strategic planning (which "centres on the setting of long-term organisational objectives and the development and implementation of plans designed to achieve them" [Stonehouse, 2002]) is a means of both reacting and adapting to variability in the manufacturing environment. It provides an approach for defining an organisation's future within the context of environmental change, so that the organisation can operate on a more effective and efficient basis. Strategic planning differs immensely from long range planning. In long range planning, a plan is developed by extrapolating on the past in order to predict for the future. However, a major drawback of such a plan is the fact that it ignores the dynamic nature of the manufacturing environment. Strategic planning, meanwhile, is concerned with addressing the organisation's mission, and defining how the organisation is to successfully anticipate and then respond to changes in the environment. The strategic planning process is not executed in isolation. It involves specialists within the organisation, concerned with the development of an overall governing framework - a framework in which the organisation's business units will operate. Within such a framework, the development and commitment of the separate business units to their own strategic plan is supported.

The overall strategic business plan is really concerned with the establishment of an organisational direction. Thus, an organisation which encourages strategic planning can be conceptually represented as a focused organisation [Reading 1993] (Figure 1.7). By allowing everyone in the organisation to be familiar with the aim of the strategic plan, organisational direction and motivation are supported.

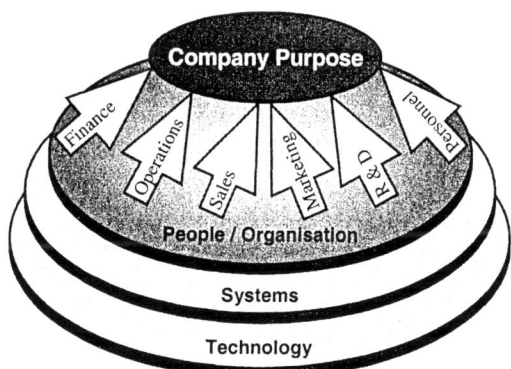

Figure 1.7 The focused organisation [after Reading 1993]

Organisations engage in strategic planning for some of the following reasons:
- If an organisation relies on and is governed by what it did in the past, rather than identifying what it wants to achieve in the future and then planning for that future, it soon loses focus in a rapidly changing world. Such a state can eventually lead to the economic death of the organisation. Paradoxically, the more difficult it is to plan, the more critical clear planning becomes.
- Corporate strategy of the 1960s and the 1970s was generally characterised by investments in comprehensive ten-year corporate planning documents, which even then soon became outdated. This was mainly due to man's inability to accurately predict the future. What is needed are strategic plans which support management in clarifying their long and short term goals, and then facilitate them in exploring potential future opportunities.

- Today's pace of change intensifies the need for those engaged in planning to maintain contact with the daily realities and directions which will impact the organisation. However, as strategic thinking is concerned with keeping in touch with such realities, the organisation is better prepared to exploit the changing environment.
- Plans frequently degenerate into a list of organisational aspirations and wishes which would be feasible in an ideal world. However, in today's less than ideal world, strategic planning deals with current realities and considers the organisation in its present position. "One needs to ensure that the plan is realistic and sets out clear steps for its implementation. Too often business plans only consist of catchy mission statements without any real evidence that the organisation has worked out how to move forward" [Lawrie 1996].

The activities involved in the strategic planning process include:
- **Statement of the organisational mission.** The mission statement provides a statement of the purposes, goals, beliefs and ethics of the organisation. Whilst it does not deal with the *what, how* and *when* of future activities, it does represent a long term statement of intent. This statement of intent follows on from the original vision which inspired the setting up of the organisation. It basically represents the broadest possible portrait of the organisation's vision of its own future.
- **Identification of the strategic aims.** The strategic aims embody the key organisational priorities and direction in the immediate to medium term future (an example of a strategic objective is the need to increase the number of customer orders by 10 percent). Whilst all organisational activities must be related back to a strategic aim, the key characteristics of strategic aims include the following :
  o Strategic aims present an external focus in the sense that they describe the outcomes or effects of the activities of the organisation on its customers.
  o Measurable by nature, strategic aims describe the minimum acceptable level of organisational performance.
  o Because they are achievable, strategic aims challenge the organisation to perform better.
  o Strategic aims are explicit and easily understood. Thus, individuals who are not familiar with the organisation will be able to understand the anticipated outcomes of the organisational activity from reading the strategic aims.
  o Strategic aims are extensive, in the sense that each of the anticipated outcomes of every function that the organisation performs is described in the aims.
  o Strategic aims advance the organisation in the achievement of its goals, as indicated in the mission statement.
- **Definition of the operational objectives**. The operational objectives represent detailed, costed and timed action plans, which are used to support the organisation in implementing and achieving each of its strategic aims. A mnemonic which helps the definition of operational objectives is **SMART** (Specific, Measurable, Attainable, Realistic and Timed). If the strategic aim is to reduce the number of accidents by 20 percent from 2004 levels, then the

operational objective is percentage reduction in the number of accidents from 2004 levels.

■ **Clarification of the critical success factors (CSFs).** The critical success factors are crucial to the successful achievement of the organisational aims and objectives (i.e. its strategic aims and objectives). CSFs represent the factors which the organisation believes it has to get right if it is to satisfy its strategic plan. They generally represent a mixture of hard elements (i.e. that which is tangible and easy to measure, such as output) and softer issues (e.g. processes, working culture and styles of work).

It is important that the corporation, the customer and the competition - the 3Cs - are considered in the creation of any strategic plan (Figure 1.8). Whilst each element of the 3Cs possesses its own interests and goals, collectively, they are referred to as the strategic triangle.

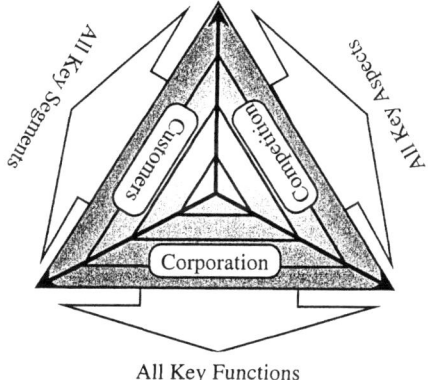

All Key Functions

Figure 1.8 The strategic 3Cs [after Ohmae 1982]

In the context of the strategic triangle, the goal of the organisation is to positively distinguish itself from its competitors, using its corporate strengths as a means of better meeting customer requirements (Figure 1.9).

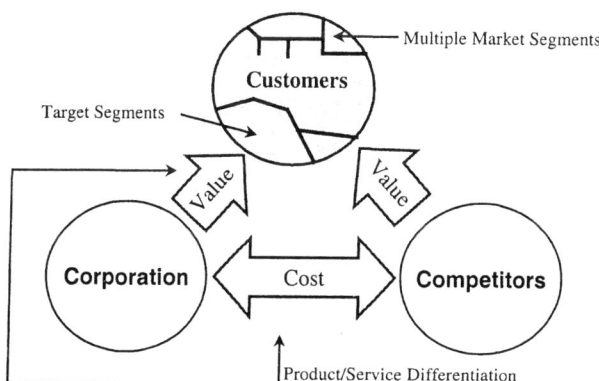

Figure 1.9 The corporation, the customer and the competition [Ohmae 1982]

Pfeifer *et al.* suggest that "the satisfaction of customer's individual desires is now, more than ever before, the key to recapturing and sustaining the competitive advantage in a market characterised by cut throat competition and customer-orientation" [Pfeifer *et al.* 1994]. Thus, it is critical that the strategic plan accurately balances the strengths of the organisation with the needs of a clearly defined market. An effective and positive agreement between organisational objectives and market demands is essential for organisational prosperity.

However, successful strategic planning is not solely about matching organisational goals to market needs, it is about ensuring a better or stronger matching of corporate strengths to customer needs than that provided by the competition. In such a matching, it is vital that management have a thorough understanding of potential customers (their needs and objectives and their geographic and demographic distribution) and competitors (their behaviour and relative strengths and weaknesses). The more an organisation understands its competitors, then the greater the differences in the approaches adopted by that organisation and its competitors. If the contrary occurs, then the result is greater customer difficulty in distinguishing between the similarly respective offerings of the organisation and its competitors. A price war might then ensue, which would invariably introduce short term benefits to the customer whilst intensifying the complexity of the management task for both the corporation and its competitors.

Traditionally, little attention was devoted to strategy by any level of management. The lower levels of management generally had no contribution to the process of strategy formulation, whilst the executives were more concerned with operational and tactical issues (Figure 1.10). In essence, this Figure suggests that all layers of management should be more involved in strategic activities.

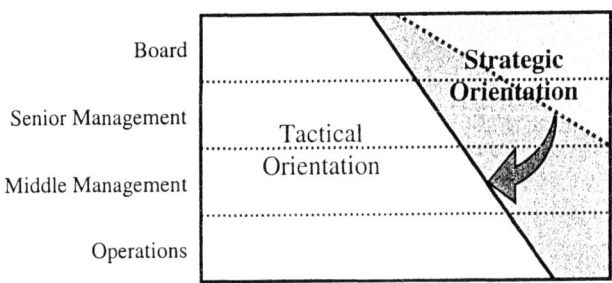

Figure 1.10 Evolution of management's role in strategy formulation

In synopsis, a strategic plan is composed of the organisational strategy and the time related details for carrying out that strategy. It also includes information concerning the marketing mix, anticipated sales and profits and control procedures to ensure that the plan is following the prescribed course. However, irrespective of the perfection of the formulated strategic plan, a critical aspect of strategic planning is timing. The most brilliant strategy will be redundant if the changeable nature of the market is ignored. Consequently, the key to organisational prosperity is dependant on both the development of a strategy that will provide the organisation with a competitive edge and the implementation of this strategy at precisely the right moment. In achieving this objective, what is required is a strategic thinker, who has the intellectual flexibility, insight and timing to support the rapid and successful organisational adaptation to changing circumstances.

The Strategic Thinker

As organisations engage in strategic planning as a means of gaining competitive advantage, there is a need for a thinking style which differs radically from the traditional approach i.e. mechanical systems thinking. A strategic thinker is characterised as having an intellectual elasticity, which endorses the rapid and successful adaptation to changing circumstances. In the strategic thinking process, a clear understanding of each element of a situation is initially developed, followed by a creative restructuring of these elements in the most effective way possible. Strategic thinking differs from both the mechanical systems approach based on linear thinking and the intuitive approach, which frequently lacks any real analysis (Figure 1.11).

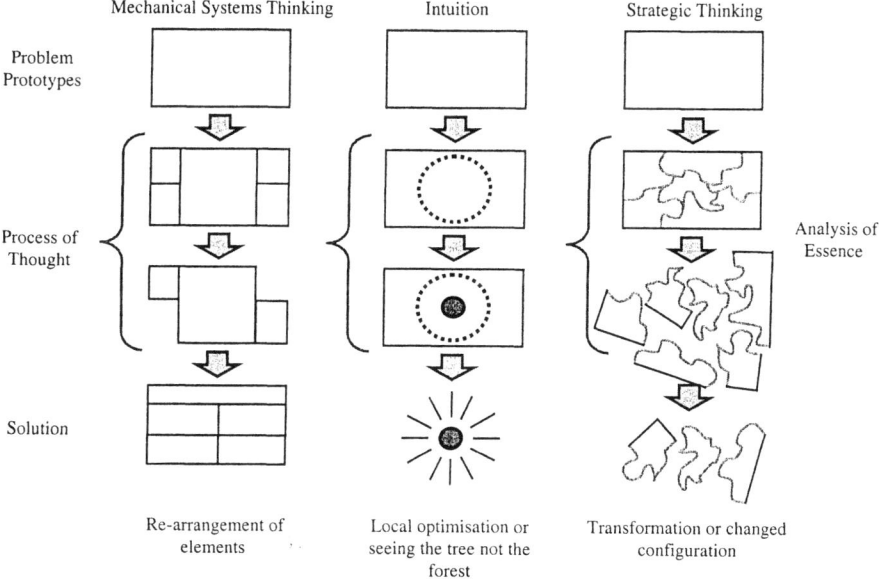

Figure 1.11 Three kinds of thinking process [Ohmae 1982]

In strategic thinking, the manager must initially identify the critical issues of a situation. Failure to successfully identify these issues will generally lead to a solution which does not adequately address the problem. Solution oriented questions can be formulated only if the critical issues are first localised and then accurately understood. Following an understanding of such issues, pressure is exerted on the manager to think creatively. It often happens that the more intense the pressure, the greater the convergence of the mental vision of the manager. Increasingly, he may rely on previously formed perceptual and/or cognitive sets. A perceptual set results when the learned conventions of an individual prevent him from perceiving the world in new ways. Meanwhile, a cognitive set results when a restricted range of solution approaches repeatedly used by an individual, detract him from the consideration of new approaches and strategies. Both are prohibitive to creative thinking and the generation of creative solutions. Frequently the differentiating success factor between competitor organisations is the strategic thinker. As he generally understands all possible alternatives and is constantly engaged in identifying the costs and benefits of each alternative, he is better equipped to respond flexibly to the inevitable changes which the organisation will encounter. It is this very flexibility which increases an organisation's likelihood of responding successfully to change.

In addition to surveying all of the organisation's critical functions, the strategic thinker must also be able to evaluate the competition in its totality, taking into account critical elements such as the competitor's research and development capabilities and its shared resources etc. The strategic thinker must be adept at mentally placing himself in the mind of the strategic planner within the competitor organisation, in order to "ferret out the key perceptions and assumptions on which the competitors' strategy is based" [Ohmae 1982]. A strategic thinker must have the capacity to "comprehend world trends as they affect business, government and standards of competition" [Stanek 2000]. Once this is accomplished, he must be skilful at timely strategic decision making. Harris likens the strategic thinker to a global leader and claims that in addition to being flexible and open, such a leader will above all develop "skills for planning and managing accelerating change as well as for dealing with knowledge workers" [Harris 2002].

Managerial Decision Making

Managerial activities are often characterised as being fast paced, unpredictable and filled with interruptions and fragmented action. A study of managers who made consistently successful decisions, identified a discernible pattern in their decision making approach. The results illustrated that in successful managerial decision making, the following conditions were met:

- The business domain was clearly and succinctly defined.
- The business environmental forces were extrapolated in order to determine possible future events. Following this, concise and logical hypotheses for the most probable of these scenarios were developed.
- On identifying the strategic route for the organisation, organisational resources (i.e. people, technology and money) were boldly and assertively deployed.
- The organisational strategy was paced according to its resources and safeguarded against trying to achieve too much too soon, and thus becoming over-reached.
- As long as the basic assumptions underlying the original strategic choice remained true, management complied with the organisational strategy. However, if changed conditions merited a modification of even the most fundamental direction of the organisation, then management would be prepared to instigate change.

Whilst the daily tasks and decisions performed by a manager are varied and intense, the managerial task has become significantly more complex due to the changing nature of time within the organisational context, where time to respond is critical and time to implement is critical. Managers no longer have the luxury of deliberating alternatives prior to decision making. The nature of today's organisational environment prohibits the testing of theories and the evaluation of conflicting ideas. The time frame allowed for managerial decisions has become increasingly compressed, whilst the need for a quicker response has become increasingly critical (Figure 1.12).

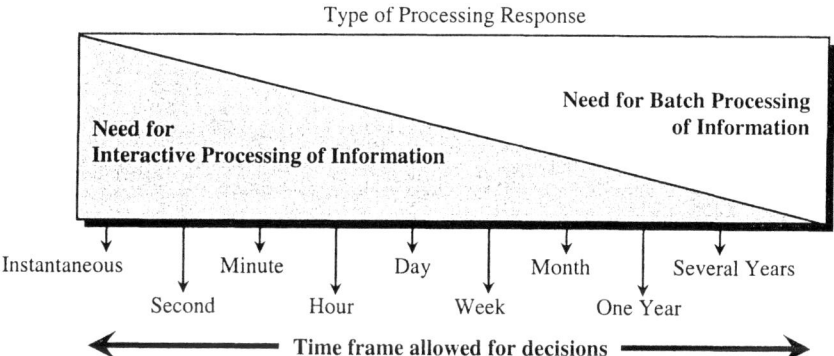

Figure 1.12 Increased interactive responses and correspondingly reduced batch responses [after Davis 1990]

The variety of decisions which are made can generally be classified under the following decision categories [Elshaug *et al.* 1995]:
- Unconscious decisions are made intuitively to maximise personal well being (e.g. a baby cries when it is hungry).
- When a decision becomes a habit, it is called a conscious decision. Conscious decisions minimise uncertainty, speed up the decision making process and reduce the need for expertise (e.g. driving a car).
- Process decisions occur when a decision is complex in nature (e.g. a user is required to select from a set of alternatives).

Within the context of process decisions (i.e. decisions which are complex in nature) the ability to reach a correct decision when conflicting and doubtful criteria prevail requires judgement. Johnson-Laird argues that judgement is accomplished through the construction of an internal representation of the external world i.e. a mental representation [Johnson-Laird 1993]. Indeed, judgement is "*more* eagerly sought and more highly paid than any ... it is an elusive quality, easier to recognise than to define, easier to define than to teach" [Vickers 1984]. Managers are commonly required to exercise judgement in decision making, in environments where time pressures, overlapping tasks and problem mutability add increased complexity to the decision making task. With the need for increasing managerial competence at rapid decision making within ever-collapsing time frames, the questioning of the rational decision making model has led to the non-rational decision making model.

The rational decision making model operates on the basis that problems can be/are well defined and that decision makers have full knowledge of the problem area and time is not critical [Travica *et al.* 1996]. The non-rational decision making model meanwhile recommends that decision makers:

- Concentrate on part of the problem without losing the holistic view.
- Use the trial and error approach in decision making.
- Access generalised information on a large body of facts and detailed information on components of a problem.

The non-rational decision making model takes into consideration the following:

- There is never sufficient information available which will readily support either an extensive analysis of the internal and external manufacturing environment or a complete exploration of alternative strategies.
- Although the available information is too limited to support an exhaustive evaluation, it is nevertheless far too abundant for decision makers to comprehend much more than a very limited and simplified set of inter-related variables. As a result, decision makers are increasingly demanding tools to filter this information and present it in forms which will accommodate and optimise managerial decision making.
- The logical, methodical and systematic approach to the planning activity can seriously impede the production of radical, but potentially successful ideas.

Whilst access to information is undeniably critical, Herbig and Kramer suggest that too much information can have negative effects on the decision making process i.e. it can result in adverse judgmental decision making. Our "limited processing capacity can become cognitively overloaded if we attempt to process 'too much' information in a limited time" [Herbig *et al.* 1994]. Information overload (which can result in cognitive strain, confusion and other dysfunctional consequences) impinges on the decision making process by reason of the following [Hedelin *et al.* 2002]:

- The decision maker is incapable of finding the relevant information.
- The decision maker overlooks that which is most vital amidst the mass of relevant information.

Thus, given that managers are frequently overloaded by information and have at times to make educated guesses concerning information validity [Melogza *et al.* 2002], what is required are tools and methodologies which support the ready access to relevant, valid and timely information so that managers are supported in making informed and effective decisions.

1.3.3 Strategy and Information Technology

Technology can be viewed as being dichotomous in nature. On the one hand, it can represent the governing force of increasing employment and competitive ability, and on the other hand, it can embody the source of major economic disruption. However, notwithstanding the fact that technology can introduce a significant amount of uncertainty into the management equation, successful management practices balance the intrinsic uncertainty of technology against the potential strategic benefits that technology can yield. As has already been outlined, strategic planning is described as a way of providing the organisation with the means to adapt its services and activities, such that the organisation can address and respond to the changing requirements in its environment. Consequently, when the strategic direction or plan is incorrect, the stability of the organisation can suffer.

Strategy and technology represent two very complex domains. Thus, in integrating strategy and technology, it is important to explore and understand strategy-technology differences and discrepancies, and to have a carefully constructed and well understood position on the nature of strategy and technology. Lack of integration between strategy and technology often results in the wasteful use of resources [Goodman *et al.* 1994]. On the other hand, the successful integration of strategy and technology, can lead to better organisational performance and increased profitability. Some of the issues arising with the integration of strategy and technology include:

- **Technology directly impacts organisations.** As production decisions generally have notable technological bases, decisions concerning process innovation and/or the reduction of manufacturing costs will have direct and significant impacts on an organisation's competitive position. Thus, as the definition of an organisation's technology affects its ability to maintain competitive advantage, technology is pivotal to market success.
- **Technology decisions are frequently strategic in nature.** Decisions concerning the implementation of certain technological choices often result in serious organisational implications. Many of these decisions can extend over long time frames and deploy large portions of an organisation's resources. Thus, whilst technological decisions are often difficult to alter once commitments are made, if they do not contribute to the success of the organisation or if they impede organisational flexibility, then they can adversely affect the corporate strategy.
- **Technology can compete for resources within the organisation.** It is important that organisations identify the optimum manner in which its resources may be deployed so that available opportunities can be exploited to their fullest. In resource deployment, organisations are required to bear in mind that the selection of one investment area normally constrains investment in another area. Thus, prior to the final deployment of its resources, an organisation should first array the technological investment projects alongside other opportunities or demands within the organisation. An assessment approach should then be used, to evaluate the long-term organisational implications of the mix of proposed activities. A lack of integration between these proposed activities could result in increased competitive risk.
- **Information systems for the management of technology are expensive.** Technology functions are often characterised by high task uncertainty and an intense need for information to be collected, processed and transferred during task performance. Thus, the management of task uncertainty and the development of appropriate information systems for the support of the technology functions have strategic implications. These implications are not only because of choices of substance and content, but also because of the costs involved.

- **The integration between strategy and technology is a governing need.** "In many organisations, strategists and technologists exist as two separate communities. Ideally, these communities maintain close communication on matters of planning and decision making. Rarely are the two groups well integrated - differences in temperament, in training, in location within the corporate hierarchy, in roles regarding budgeting, all detract from integration. These factors reduce the likelihood of an effective strategic-technological integration without purposeful, if not formal measures imposed by management" [Goodman *et al.* 1994].

1.3.4 Knowledge

The majority of today's workers (in the developed world) are knowledge workers: bankers, accountants, computer programmers, stockbrokers and market analysts. They are knowledge workers in the sense that they are primarily involved in the creation, distribution and use of information. Because of this focus on knowledge workers, today's economies are being increasingly established on the production, management and use of information.

Considered equivalent with competitive advantage [Nonaka *et al.* 1995], knowledge is recognised as one of the most valuable resources within every domain of human life. "The basic economic resource is no longer capital, nor natural resources, nor labour. It is and will be knowledge" [Drucker 1994]. However, knowledge still remains an elusive resource difficult to capture, store, ship and use at a global level [Chira *et al.* 2003]. Often knowledge is internalised in an individual or indeed a group and the sharing of this knowledge depends on a complex set of sociological and psychological factors. Moreover, the word *knowledge* is often used indiscriminately, covering meanings such as data and information, or data/information/knowledge hierarchy. Practically speaking, data is something less than information and information is something less than knowledge. Data generally represents simple facts or individual entities, which organised and structured in a meaningful context generate information (or "a flow of messages" [Nonaka *et al.* 1995]). Thus, information may be defined as data, which is used by the recipient to enhance what he already knows. It adds to knowledge and in ideal situations reduces uncertainty (when presented in a format which is both relevant and understandable). Information is characterised by its completeness, accuracy, frequency, relevance, origin, timeliness and detail. A deficiency in any one of these attributes can result in general information devaluation.

In organisational terms, prompt access to accurate and relevant information (for the purpose of solving problems and assisting in strategic decision making) is useful in gaining a competitive edge. Meanwhile, the absence of critical information can result in erroneous decisions, neglected opportunities and serious performance related problems. As a result, it is important that managers have a means of acquiring relevant and timely information in order to accommodate their needs.

The analysis, synthesis and interpretation of information (Figure 1.13) create meaning and therefore knowledge [Tuomi 1999, Shaw 2003, Srinivas 2003].

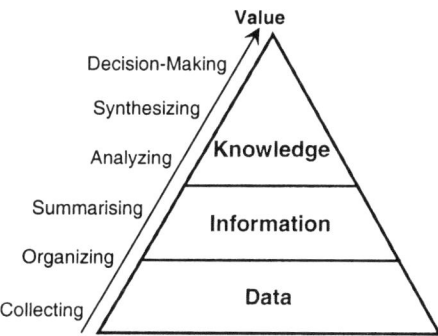

Figure 1.13 Data, information and knowledge value chain [Srinivas 2003]

The management of data implies the processing of large amounts of simple facts, while the management of information translates these facts into meaningful frameworks [Shaw 2003]. Usually, these processes do not require human intervention and therefore the systems that implement them are fairly automated, robust and effective. Conversely, the process of knowledge management requires human participation (after all, the human is the final destination - consumer - of any knowledge related process).

Therefore, knowledge must be organised and managed so that human access to it is supported. In any equation that involves knowledge, the human is also a factor to be considered and vice-versa. The philosopher Michael Polanyi and the Japanese organisation-learning theorist Ikijuro Nonaka suggest that knowledge has two forms i.e. implicit knowledge and explicit knowledge [Polanyi 1966, Nonaka *et al.* 1995, Nonaka *et al.* 1998]. Implicit or tacit knowledge represents personal knowledge stored in the bearer's mental structures, having its roots in the private psychological baggage of the individual (e.g. subjective insights, intuitions and hunches). This kind of knowledge cannot be easily formalised, hence it cannot be straightforwardly communicated (at least not by the means of a formal language) or shared. *Explicit knowledge* is the knowledge codified and systematically expressed in formal structures compatible with human language (e.g. libraries, archives, databases). Hence, the explicit knowledge represents the kind of knowledge that is communicated and shared.

The process of managing explicit knowledge consists of the following [Shaw 2003, Sirinivas 2003, Benjamins *et al.* 1998, Sure *et al.* 2002].
- **Knowledge gathering** (capturing): acquisition and collection of knowledge.
- **Knowledge organisation** (structuring, storing): structuring the knowledge acquired in order to manage it effectively.
- **Knowledge refinement**: correcting, updating, adding and deleting knowledge.
- **Knowledge distribution** (sharing, dissemination): pass knowledge to the professionals who need it.
- **Knowledge using**: employing knowledge in the decision making process.

An important aspect of knowledge management is the process of generating new knowledge. Nonaka *et al.* advocate an organisational management approach towards knowledge creation and management.

Nonaka and Tagushi [Nonaka *et al.* 1995] have identified four types of interaction that can generate knowledge (see Table 1). *Socialization* (from tacit to tacit) is the process of sharing experiences, mental states, and technical skills within groups or organisational cultures. The key element of socialization is considered to be the person's experience. *Combination* (from explicit to explicit) generates new explicit knowledge through activities such as sorting, adding, combining and categorizing existing explicit knowledge. *Externalisation* (from tacit to explicit) creates explicit knowledge through the use of metaphors, analogies and models. *Internalisation* (from explicit to tacit) supports learning by doing.

Table 1.3 The four modes of knowledge conversion (after Nonaka *et al.* 1995)

From \ To	Tacit knowledge	Explicit knowledge
Tacit knowledge	Socialisation	Externalisation
Explicit knowledge	Internalisation	Combination

It is widely believed that "we can know more than we can tell" [Polanyi 1966], and therefore the aim of a complete knowledge management system should be twofold as follows: firstly, to manage effectively the explicit knowledge and secondly, to support, promote and enable human's implicit knowledge, as this will affect creativity.

1.3.5 Managerial Need for IT based Tools and Technologies

No manager can be expected to systematically follow the complex task of information processing, which opens up a myriad new alternatives, corresponding to as many as possible scenarios, and thereby further amplifies the already exacting number of interrelations and their consequent effects on overall objectives.   – Pratali 2003

The relentless progress in information technology has resulted in a short-circuit of time, space and processes, with information being relayed around the world at very high speeds. Management is realising that the proactive use of information and information technology is critical to organisational success. Whilst information technology (IT) impacts the entire nature of the organisation from its design and structure to its strategies, the speed with which management receive information is pivotal to the value adding nature of information and IT. IT, by supporting the rapid and timely access to relevant and accurate information, also supports management in making decisions that they might otherwise have been averse to making or unable to make.

The information system support of organisational performance can result in the gains in productivity, competitive advantages and overall increase in organisational effectiveness. The manager as the activating element within the organisation increasingly requires access to information systems which will support him in his managerial role. Managers use information systems for many reasons. See Table 1.4.

Table 1.4 Why managers use information systems [after Senn 1990]

Reason	Explanation
To keep informed	Allows the manager to keep apprised of activities within the organisation, without getting bogged down in a forest of minutiae.
To analyse data	Permits the examination of alternative scenarios and supports 'what if' analysis.
To browse through data	Allows the executive to browse through the data first-hand. As a result, he can attain a view which cannot be achieved through a review of reports of business summaries.
To understand a new situation quickly	Provides the manager with rapid access to business details.
To maintain surveillance	Supports the monitoring of a "situation of specialised interest through specified details" [Senn 1990].
To get at data directly	Enables the manager to get directly to the data he requires, thus eliminating the need to wait for staff to retrieve and extract the necessary details.

The more serviceable and useful the available information, then the more valuable it in turn becomes. However, whilst information is increasingly accessible within an ever-collapsing time frame, Hanka argues that management may soon reach the point where they cannot cope with the increased knowledge available to them [Hanka 2000] (Figure 1.14), unless information(/knowledge) systems are available which support the efficient and effective management, filtration, searching and presentation of that information and knowledge.

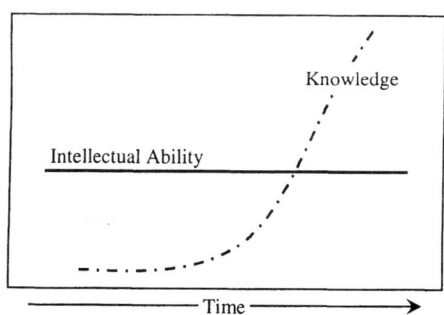

Figure 1.14 Intellectual ability vs. knowledge [Hanka 2000]

Information systems are human activity systems. They comprise information, processes and human participation, in the support of organisational performance by:
- Creating new channels of information for management.
- Filtering incoming information.
- Supporting management in experimentation and what if analysis. This indirectly results in the expansion of the thinking ability of the brain of the organisation (i.e. the change master).

Information systems should ideally emphasise the value of the information and promote the worth of such information as a strategic commodity to the organisation. From this perspective a successful information system should:
- Improve the manager's conceptual model of the organisation.
- Enhance the thinking approach adopted by the manager.
- Foster enhanced planning and control capabilities.
- Support the manager in optimising his time.
- Illustrate the potential to be reaped from the successful use of information technologies.

Managerial information needs might thus include the following:
- **Comfort information** pertains to information aimed at keeping the manager apprised of the present position e.g. last month's sales figures.
- **Externally distributed information** represents information released to stockholders and the media e.g. quarterly corporate earnings report.
- **Planning information** represents descriptions of future developments and plans e.g. anticipated growth or shrinking of the organisation's market.
- **Internal operations information** refers to organisational and employee progress, evaluated on the basis of key factors or parameters e.g. percentage share of market held by the organisation.
- **Status or progress information** ensures that the manager maintains contact with current problems and crises e.g. a competitor's progress on a similar product.
- **External intelligence** represents a type of soft information, which encompasses gossip, opinions and expected shifts etc.
- **Warning information** informs the manager of impending changes e.g. expected price fluctuations.

Theoretically, information systems in the form of EISs (executive information systems) and DSSs (decision support systems) are truly beneficial to managers. However, the reality is that managers are not using and exploiting them to the extent anticipated. Studies indicate that managers generally perceive barriers to information systems use and deficiencies in information systems design. The greatest obstacles to information system use, seem to include the following:
- An inadequate graphical user interface. "Sometimes, software seems so mind-bogglingly stupid that it's almost impossible to believe that any human being other than the programmer actually worked with the product before its release" [Randall 1998]. This view is also supported by Constantine who states that "all the major software houses have elaborate user interface testing programmes supported by expensive usability labs. We lavish more and more attention on

user interface programming – and still miss the mark by miles" [Constantine 1998].

- The rigid and structured nature of the data complicates experimentation and what if analysis.
- The absence of input by top managers to systems design.

It is apparent that these obstacles remain today, for Stonehouse *et al.* surveyed 159 small and medium sized enterprises selected from both the service and manufacturing sectors with respect to the use of strategic frameworks and tools of strategic analysis [Stonehouse *et al.* 2002]. Their findings showed that few of the organisations surveyed use traditional tools of strategic management (i.e. STEP, the five forces framework, value chain analysis and portfolio analyses) and of strategic planning.

The results of a survey conducted by the Market Research Institute[3], with respect to managerial technological literacy stressed the importance of effective user interface design [Market Research Institute 1995]. This study indicated that ninety-one percent of the senior managers interviewed, agreed that despite the increasing importance of managerial technological literacy, less than half of their organisations' senior managers were technologically literate. The relationship between managerial technological literacy and organisational competitiveness was emphasised by the interviewees [Market Research Institute 1995]. Likewise, a user survey taken at an EIS conference illustrated that the main priority of executive users are systems which promote ease of use. This survey stressed that more focus is required at the managerial computer interaction level [The Economist Intelligence Unit 1991]. Given the increasing attention to strategic management, there is clearly an overwhelming need for design tools which cater to the representation of individual user cognitive system models i.e. systems which meet the targeted user's needs.

As indicated earlier managers, particularly senior managers are generally intermittent users. Due to such intermittent use, they tend to forget how to operate the system from one interaction to the next. Consequently, they must be able to see some inherent value in using the information system in the first place. More often than not, managerial users tend to unconsciously assign checks as they navigate and interact with the system. These checks describe the effort required in using the system (in order to achieve value). In general, managers have a lower tolerance threshold than most people, to badly designed systems. They are just not able to spend the time which is needed to learn the use of a difficult system. Thus, it is vital that systems be designed which meet their needs and are user friendly.

In an ideal setting one could liken an information system to a crystal prism [Boone 1991]. If a crystal prism is rotated in the path of a light source, a different colour is reflected in each facet of the prism. However, if the prism is secured to a desk, then naturally, it becomes difficult to examine it from all viewpoints. Information on

---

[3] Although a few of the studies reported here are somewhat dated the authors believe that the results reported are still valid.

paper hinders viewing from different perspectives, in the same way that an anchored prism hinders viewing, for once ideas are committed to paper they become static. Once they are static, they lose their multiple perspectives, for they can no longer be easily examined from all angles. They can no longer be changed or re-arranged. Consequently, just as a crystal reflects different colours when viewed at different angles, the information system's support of multiple information views creates the possibility for an information prism.

The term information prism is used to describe the manner in which an information system can expand the boundaries on information. This type of expansion represents the main difference between information systems and paper based static reports. The information system's support of information experimentation, alteration and re-arrangement represents the main reason why it is so important that executives work in a hands-on capacity with information systems, in order to fully benefit from their use. "If someone else operates the computer, the executive loses the speed and flexibility with which he can formulate change and work with thoughts and ideas. If information tools can help executives think, they can help corporations flourish. Computers in and of themselves are worth very little. They deliver value only when the right tools are applied in the right way to a specific business challenge" [Boone 1991]. However, despite the fact that managers may indeed have the correct information systems at their disposal, it does not guarantee that such systems will be used. As Strong suggests, if the "interface is ineffective, then the system's functionality and usefulness are limited, users become confused, frustrated and annoyed, developers lose credibility and the organisation is saddled with high support costs and low productivity" [Strong 1994]. Because managers are generally intermittent users, the design of the graphical user interface (GUI) needs to reflect this state.

1.4   Conclusions

With the recent developments in information and associated technologies, computers and communication-related media have pervaded virtually every aspect of modern life, resulting in short-circuit of time, space and process. This information technology revolution has influenced the shrinking of international borders and the creation of global markets. If organisations are to be competitive, their business activity must now take place within this global setting. The emergence of the global market has seriously impacted the manufacturing domain. Product life cycles are becoming progressively shorter. The incorporation of technological innovations into new products is required within even shorter periods of time and the requirements to be met by both products and services are increasingly high. Managers are increasingly required to make effective strategic decisions within ever collapsing time frames. The socio-economic and socio-environmental context within which organisations and managers function has changed with an increasing consideration of issues which in the past might have been considered to be outside of the scope of management. Essentially increasing competition in the business world and changing social and legislative contexts have meant that despite the fact that our view into the future has never been less clear, the necessity of a cohesive organisational strategy and plan going forward has never been greater. There is a pronounced need for managerial systems which encourage and support the task of strategic planning and decision making.

The successful organisation is one which is flexible by nature and adept at delivering rapid response to changing market opportunities. Today's organisations must be capable of responding to change and developing and implementing business strategies which are responsive. Information Technology and Information Systems are essential tools in support of the strategic planning process and when implemented properly support and facilitate good decision making by management. Although there has been significant effort expended in strategic analysis tools, the evidence suggests that managers are not making sufficient use of such tools. Inadequate design in terms of the matching of functionality and presentation to the needs of managers, over rigid data structures and inadequate user interfaces are often to blame for the relative failure of such systems.

> IT is of course an essential tool for the modern world, a prerequisite for economic competitiveness in so many areas where speed of communication is half the market battle, and indeed for improving the quality of life in other areas. The problem is not IT itself. It is the human tendency to take the soft option by reducing everything else to IT. If individual creativity is to be confined within the frameworks permitted by IT, then however wonderful the technological advances, there will be a drying up of part of the human potential.
>
> IT offers enormous educational opportunities ... but if used unimaginatively, it can pose serious dangers...
>
> There is a paradox in that when we turn to IT to ensure our competitiveness in the international economy, we may be creating a new type of dependency culture in place of an old one, symbolised at the most elementary level by calculating machines substituting for mental arithmetic and spell checks for spelling. Far from expanding the potential of the mind, this type of dependency on IT actually shrivels it.
>
> - Lee 1998

Chapter 2

Performance Measurement

Keywords: Performance Measurement, Traditional performance measurement systems, contemporary performance measurement systems, performance measures, TOP, TOPP, ECOGRAI, Balanced Scorecard, PMQ, business processes.

Chapter Objectives: All high-performance organisations, are, and must be, interested in developing and using effective performance measurement and performance measurement systems, as it is only through such systems that these organisations can maintain their high-performance. The purpose of this chapter is to introduce the reader to performance measurement and some of the performance measurement systems which are currently available. Having read this chapter, the reader should become familiar with:

- The Performance Measurement Process.
- The difference between traditional performance measurement and contemporary performance measurement systems.
- The TOP approach.
- The TOPP approach.
- The ECOGRAI approach.
- The PMQ approach.

2.1 Introduction

> We live in a measurement culture, largely as a result of globalisation.
> – Line 2003

An organisation's ability to survive and function successfully partly depends on the information available to its management. The role of information relating to performance, whether for use within the organisation or for external stakeholders, is a key factor in determining commercial viability. Concerns are now being raised in the business environment over how best to measure an organisation's performance. This chapter introduces performance measurement in an organisational setting. The concepts discussed can be applied to aspects in the organisation from the highest level down to the area where a specific task is accomplished.

Performance measurement is the trigger for performance improvement and the statement "If you don't measure it, you can't improve it" very often holds to be true. Organisations tend to use a wide range of measures to gauge their overall performance. These measures are seen by management and indeed by all employees, to be important since they reflect what the organisation believes to be important. If something is not important, then why would it be measured? and why would attempts be made to either increase or decrease the value of the performance measure? Thus, the performance measurement system used by an organisation not only provides an explicit reminder of how the organisation is performing but it also implicitly shapes the actions of employees. As a result, it is important that the performance measures being collected by an organisation are reflective of the strategy of the organisation and/or the requirements of the organisation's customers. It should be noted however, that whilst organisations tend to collect large numbers of performance measures (PMs), they generally tend to use only a very small subset of these measures to make decisions.

Given the importance of organisational processes, it is becoming increasingly important that organisations adopt effective performance measurement systems as the normal way of working, in order to support the continuous improvement of an organisational process or goal [AEDC 1996]. Cotta-Schonberg argues that whilst the terms performance measurement and evaluation are often interchanged, they are fundamentally dissimilar. He believes that it is important to "keep description of which measurement forms a part, distinct from the value judgement, which is evaluation" [Cotta-Schonberg 1995].

Performance measurement cannot be described as a basic problem solving approach. Whilst it may support the user in resolving a particular problem, it does not provide the solution as such. In performance measurement, a set of measures are identified. These are used to provide an explicit representation of the performance of the organisation. Thus, performance measurement basically supports the organisation in determining whether specific goals and objectives are being accomplished. Such an approach is useful in evaluating whether or not the organisation is fulfilling its goals and whether or not such goals have a continuing relevance. As a result, it is vital that the performance measures selected by management reflect the organisational strategy. Measures compatible with the vision of the organisation foster a focused organisation (see Chapter 1, Figure 1.7). If incompatibility exists between the performance measures and the organisation's goals, then the performance measurement system will be at best ineffective or, more likely, counterproductive. Counter productivity can result in individual business units trying to improve organisational performance by doing what they perceive to be most effective for their particular function, as opposed to planning for the overall organisation.

In addition to creating the basis on which the performance of the organisation is assessed, performance measures strongly influence management in the strategic decision making task. Consequently, the importance of performance measurement as an influence on the strategic direction of the organisation cannot be ignored (Figure 2.1).

Performance results usually validate the fact that an organisation did the right things right and provide the baseline stimulus for future quality improvement activities.   – O'Leary 1996

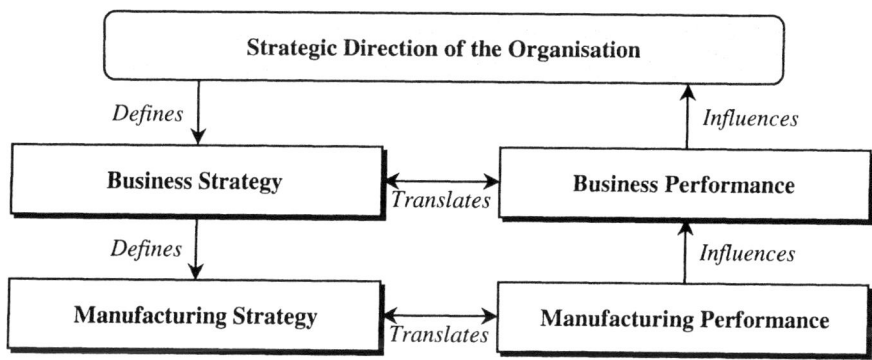

Figure 2.1 Influence of performance on the strategic direction of the organisation and vice versa

Performance measurement systems can generally be classified as being either traditional or contemporary. The former refer to approaches used by organisations having a functional, control oriented culture, whilst the latter embrace a team based business process oriented culture [Bradley 1996]. The performance measurement framework, which forms a fundamental part of the AMBIT approach, is representative of a contemporary performance measurement system. It will be shown in the following chapters that the AMBIT approach satisfies the necessary requirements for an effective performance measurement system, i.e. the need for [Chan *et al.* 2003, Kennerley 2002, Waggoner *e. al.* 1999, Barth 1996]:

- A direct relationship to the mission and objectives of the organisation.
- Validity.
- Accuracy.
- Enhancing motivation and communication.
- Completeness.
- Diagnosing problems.
- Reliability.
- Understandability and usability.
- Flexibility.
- Adaptability.
- Supporting the understanding of the organisation's progress and the current competitive position of the organisation.
- Maintainability.

It is worth noting at this point that much of the work on performance measurement carried out in the 1990s and indeed many of the performance measurement systems reported later in this chapter were designed with benchmarking in mind. The main motivation for those systems was concerned with understanding how well a particular business or more frequently a particular business process was performing compared to what were frequently termed 'best in class' or 'world class' standards. Such comparisons provided the basis for determining the potential for process improvement, which in turn led to the initiation of business process re-engineering projects.

2.2   Performance Measurement

Performance measurement is the process of measuring an activity's efficiency and effectiveness. It is a set of metrics used to assess the efficiency and effectiveness of actions. A performance measurement system is not simply concerned with collecting data associated with a predefined performance goal or standard. It is an overall management system involving prevention and detection aimed at achieving conformance of the work product or service to customer requirements. Effective performance is not merely measured by the delivery of results in one area, but by delivering satisfactory performance across all the measures. Furthermore, performance measures are intended to reflect performance as well as stimulate improvement, without being unduly oppressive. It is a continuous cycle in order to expand and improve the work process or product as better techniques are discovered and implemented.

Simply put, performance measures inform organisations about their products, services and the processes that affect them. They are tools to help organisations understand, manage and improve what they do. Furthermore, they let any organisation know:
- how well it is doing,
- where its strengths and weaknesses lie,
- if it is meeting its goals,
- if its customers are satisfied,
- if its processes are in statistical control,
- if, and where, improvements are necessary.

In short, they provide organisations with the necessary information to make educated decisions. They are recognised as an important element of all continuous improvement programmes. Performance measures do not simply describe what has happened; they influence what will happen, as they provide information for decision makers to make decisions which may affect the future competitive position of the organisation. A well designed performance measurement system is a strong tool for controlling business objectives. Measurement can be used to achieve objectives through targeting the processes that support organisational objectives. However, measuring the wrong things in the wrong area or at the wrong level in an organisation can prompt an inappropriate response and affect the ability to achieve organisational objectives. In other words, by measuring the wrong things an organisation is encouraging employees to do the wrong things. This is particularly evident if the measurement influences their pay. Such an activity will pull the organisation further away from its corporate objectives and will act as an impediment to change and improvement [Baldry et *al.* 2002].

The United States Department of Energy [USDOE 1997] developed a 'feedback loop' that illustrates the steps for maintaining conformance to goals or standards (Figure 2.2). Conformance is achieved by communicating data back to a worker responsible for the activity and/or decision maker who will initiate the appropriate action. The key message of this loop is; those responsible for managing the critical activities must always be in a position to know the following in order to achieve the organisation's goal or standard:
- what is to be done,
- what is being done,
- when to take corrective action, and
- when to change the goal or standard.

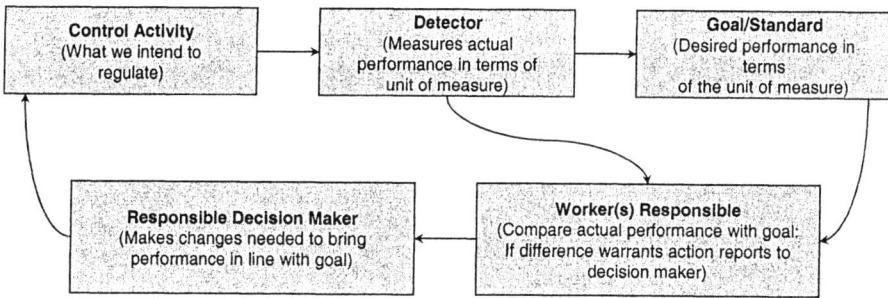

Figure 2.2 The goals feedback loop [USDOE 1997]

The basic elements of the feedback loop and their interrelations are as follows:
- A detector assesses the actual performance.
- He then reports this performance to an employee responsible for the process.
- This employee receives information on what the goal or standard is.
- He then compares the actual performance to the goal. If the difference warrants action, the employee responsible for the process reports to a responsible decision maker.
- The responsible decision maker verifies the variance, determines if corrective action is necessary, and if so, makes the changes needed to bring performance back in line with the goals.

2.2.1 Units of Measurement

A performance measure is composed of a number and a unit of measure. The number relates to a size (how much) and the unit gives this number a meaning (what). They are always tied to a goal or an objective (the target). They can be represented by single dimensional units like hours, meters, seconds, pounds, number of reports, number of errors etc. More often, multidimensional units of measure are used such as, kilometres per litre, number of accidents per million hours worked, number of on-time vendor deliveries per total number of vendor deliveries. These measurements are more meaningful to those who have to make decisions. An ideal unit of measure:

- reflects the customer's needs as well as the needs of the organisations,
- provides an agreed upon basis for decision making,
- is understandable,
- applies broadly,
- may be interpreted uniformly,
- is compatible with existing measures,
- is precise in interpreting the results,
- is economical to apply.

2.2.2 Performance Measurement Process Overview

USDOE [USDOE 1997] developed a high level view of the performance measurement process, which, depicted in Figure 2.3, details 11 discrete steps.

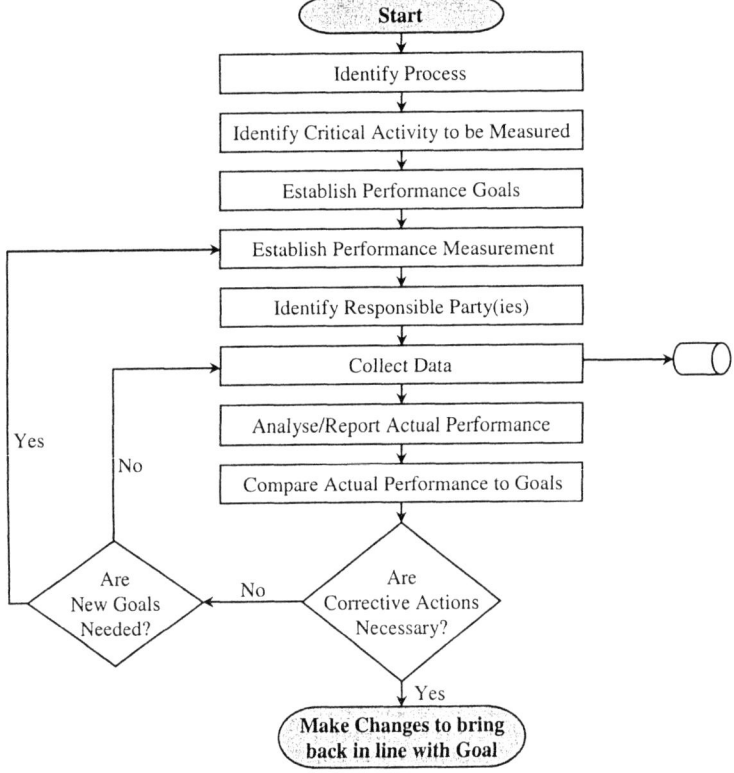

Figure 2.3 Steps in the performance measurement process [USDOE 1997]

1. **Identify the process flow**: A process must be understood in order to be controlled. A Flow diagram is a useful tool to help us understand a process. By flowcharting the entire process, the foundation for developing performance

measures is laid. All parties who are involved in the process should participate in creating the flowcharts. By doing this, individuals will receive a new understanding of the processes that affect them. Furthermore, by involving those affected, employee support (which is imperative in making the performance measurement system work) is secured.

2. **Identify the critical activity to be measured**: It is important to focus on key areas and processes as opposed to people. Each activity in the process should be examined and those activities which are recognised to be critical should be highlighted. Critical activities are those that significantly impact efficiency, effectiveness, quality, timeliness, productivity or safety. Critical activities also impact management priorities, organisational objectives and customer requirements.

3. **Establish performance goals or standards**: These goals should be the target to which you are aiming. Goals can be either set by management or they can be set in response to customer needs or complaints. They should be specific, measurable, attainable, realistic, consistent, understandable, equitable and customer oriented.

4. **Establish performance measurements**: Performance measures should, where possible, be stated in the same units as the organisation's goals. However, if an organisation has not developed goals or targets, the performance measures should measure what the organisation wants to know.

5. **Identify individual(s) responsible**: At this stage an individual or team should be assigned to take responsibility for the process. Ideally, this person or team of people will be a senior member of staff as lack of authority may prevent an individual from performing his job. This person should also be accountable for the process.

6. **Collect data**: Data are a set of facts presented in either numeric or descriptive form. Data must be specific enough to provide definite information. Furthermore, it is essential that the data which is collected is accurate and free from bias.

7. **Analyse/report actual performance**: Raw data must be analysed and assembled into a performance measurement. It should then be summarised in a graph or table.

8. **Compare actual performance to goals**: The individual or team responsible should see whether the actual performance differs to the desired performance. If there is a significant difference, those responsible must ascertain if action is necessary. There are several alternatives available for possible actions, which include; ignoring it, changing the process or changing the goal.

9. **Ascertain whether corrective action is necessary**: If the variance is large there may be a problem with the process and corrections will be needed to bring the performance back in line with the desired goal or standard. A root cause analysis may be undertaken to evaluate the potential problem.

10. **Make changes to bring back in line with goal**: All identified sources of defects should be removed and a new or improved process should be developed

11. **Ascertain whether new goals are needed**: The decision to create new performance measures or goals will depend on three major factors: (a) the degree of success in achieving previous objectives (b) the extent of any change

to the scope of the work process and (c) the adequacy of current measures to communicate improvement status relative to critical work processes.

It is suggested that these steps should be adapted where necessary to fit the specific needs of each organisation and/or process within an organisation. Material should be amended or new information added whenever it is appropriate for a particular process.

2.3   Traditional Performance Measurement Systems

Traditionally manufacturing performance measurement has been confined to cost performance, which typically drove manufacturing executives to manage the process to produce at minimum unit cost, while generating as few negative variances from standards as possible. These traditional performance measures originated from the fifteenth century when double entry bookkeeping was developed and thus have been closely linked to management accounting. Such measures were the main basis for decision making in organisations throughout the early industrialised world. In fact this remained the case until well into the late 20th century.

As global competition intensified and organisations concentrated on core competencies and movement to demand pull strategies, such as just-in-time (JIT) and efficient consumer response (ECR), organisations found that traditional performance measurement systems were inappropriate; the performance measures were not supplying organisations with the information required to compete effectively in the market place. Therefore, while cost based performance measures at one time represented the main basis for decision making in organisations throughout the industrialised world, in themselves and used alone, they have become incompatible with the business environment of today. It has been suggested that the following characteristics of traditional performance measurement systems have influenced their decline [Chan *et al.* 2003, Neely 2003, Burns 1998, Kaplan *et al.* 1992, Maskell 1991, Dixon *et al.* 1990]:

- Cost distorting.
- A general lack of relevance.
- Inflexible.
- No support for continuous improvement.
- Little consideration for customers or competitors.
- More internally than externally focused.
- Scant indication of future performance.
- Impediment to progress in modern manufacturing environment.
- Subject to the needs of financial accounting.

Traditional performance measurement systems can impede organisational progress in world class manufacturing, due to the conflict between such systems and the principle of good modern manufacturing practice, including for example small batch production, lower inventory and fast changeover times. Traditional performance measurement systems cannot account for the changes occurring in the business environment with increasingly demanding and discerning customers and more competitive markets [Neely 2003]. They are often concerned only with the efficiency at an individual production work centre, and assumed that if the efficiency of individual work centres is monitored, then the total efficiency of the plant can be maximized automatically. Furthermore, traditional performance measures cannot supply the business pertinent qualitative information required in order to compete in the business environment. As customers can now get products from a range of organisations for about the same price, other factors form the basis for competition between organisations. These factors include time and quality. A customer doesn't just want a cheap product, he also wants a quality product and he wants it quickly. This has resulted in a performance measurement revolution [Neely 1999] and has forced organisations to change the performance measures they were using and introduce new competitive dimensions, such as:

- **Effectiveness**: Is the process output conforming to customer requirements. In other words is the organisation doing the right things?
- **Efficiency**: Is the process producing the required output at minimum resource cost? In other words is the organisation doing things right?
- **Quality**: Does the product or service meet customer requirements and expectations?
- **Timelines**: Is the unit of work done correctly and on time? Criteria must be established to define what constitutes timelines for a given unit of work. It is usually based on customer requirements.
- **Productivity**: Is the value added greater than the value of the labour and capital?
- **Safety**: Measures the overall health of the organisation and the working environment of its employees.

This movement from traditional performance measurement systems to contemporary performance measurement systems is supported by the following [Zairi 1994]:

- The management approach has moved from manager centered to customer centred. The emphasis in a modern organisational context is delivering quality rather than producing quantity.
- Direct physical measures are an effective means to decision making. Unlike traditional measures, measures such as cost, quality and time can lead to action on the spot and decisions taken at the right time to make the necessary adjustments and bring about any corrections.
- Measuring the capability and the consistency of the process determines the overall capability of the organisation and as such enables senior managers to define parameters of competitiveness.
- New measures can support strategic direction and make goal setting a more achievable task. One of the fundamental attributes of an effective performance measurement system is that it should encourage actions congruent with the organisation's business strategy.
- Performance measurement has to fit the culture of the organisation.
- Performance measurement should be about managing the process and not controlling people.
- Performance measurement is evaluated through group performance.

- Modern business productivity is based on people productivity. Therefore the performance measurement system must be linked to the reward and recognition systems.

2.4   Contemporary Performance Measurement Systems

Contemporary performance measurement systems evolved in response to the realisation that traditional performance measurement systems, could not adequately address organisational needs in an environment of great change. Whilst cost historically represented the major basis of organisational competition up until the 1940s, other factors such as quality, time, flexibility and the environment are currently dictating the terms for competitive success. As a result, organisations have been compelled to derive new (or contemporary) performance measurement systems which reflect such changes in the business environment. The attributes that are characteristic of contemporary (and successful) performance measurement systems are as follows [Neely 2003, Kennerley 2002, Baldry 2002, Bourne *et al.* 2000, Burns 2000, Maskell 1991, Dixon *et al.* 1990]:

- they are directly related to the manufacturing strategy,
- they are mutually supportive and consistent with the organisation's operating goals, objectives, critical success factors and programmes,
- they convey information through as few and as simple a set of measures as possible,
- they allow managers to learn about the future,
- they support organisational employee learning,
- they monitor developments and changes in the internal organisational environment and raise action signals to indicate when set threshold levels/flags have been reached,
- they are dynamic and continue to reflect the issues of importance to the organisation,
- they provide fast feedback to managers,
- they reveal how effectively customers' needs and expectations are satisfied,
- they change over time, as needs change,
- they primarily use non-financial measures,
- they are simple and easy to use,
- they foster continuous improvement,
- they alter the organisational culture,
- they provide a set of measurements for each organisational component that allows all members of the organisation to understand how their decisions and activities affect the entire business.

Given that one of the characteristics of contemporary measures is that they change over time, in relation to changing needs, it is clear that the introduction of new performance measures should correspond with the introduction of new manufacturing techniques. However, in introducing new measures, it is important that the old measures be removed. If the new measures are used in conjunction with the old, then the new measures will not have their intended usefulness and impact. They will either be largely ignored because people are familiar with the previous methods, or both sets of measures will be used and the organisation will not gain the coherence and focus that the new measures intended to offer.

The implementation of a performance measurement hierarchy is an essential part of any manufacturing excellence programme. Meanwhile, the process by which the hierarchy is developed is crucial to the success of the implementation. The fact that the business is committing time and resources to the development process, provides motivation and credibility to the exercise and ensures that [Baldry 2002, Russell 1992]:

- "Cherry picking" is avoided. This means that the measurement system as a whole is implemented, not just one or two easy or attractive measurements picked out.
- Business processes, which are important to the success of the business, are properly measured.
- Results must be used or else they will not be taken seriously.
- The linkages and conflicts are made visible and managed, not hidden under the table. An appropriate set of measurements is used to avoid sub-optimisation.
- The measurements are stable, i.e. not changed due to internal political pressures or "flavour of the month" ideas, whilst allowing a managed evolution, as business needs change.
- When times get hard, people will be able to say "This is why we decided to do it this way. Does what we are talking about changing really make sense?"

In addition to supporting the strategy of the organisation and the viewpoint of the customer, performance measurement systems should also be process oriented as all the activities carried out in an organisation are performed as part of a process. It is the performance of this process that is ultimately of interest to the organisation as this is where value is generated and costs incurred.

Before examining performance measurement approaches, the following terms need to be defined, namely [Chan *et al.* 2003, Neely 1994, Neely *et al.* 1994]:

- A **performance measurement system** can be defined as a set of metrics used to quantify both the efficiency and effectiveness of actions. It also provides the essential management feedback for decision makers.
- **Performance measurement** can be defined as the process of quantifying the efficiency and effectiveness of action.
- A **performance measure** can be defined as a metric used to quantify the efficiency and/or effectiveness of an action.

For a given performance measure to be specified, it is necessary to define, among other things, the title of the measure, how it will be calculated/measured (the formula), what are its performance indicators, sub-processes associated with the measure, etc. For example, the process of customer order fulfilment can be described as shown in Table 2.1. If necessary, each of the sub-processes can be expanded on similar lines. An extensive study of performance measure specification for most manufacturing processes can be found in ENAPS [1999].

Table 2.1 Performance measurement specification for customer order fulfilment

Name	Customer Order Fulfilment (COF)
Description	Is a business process and includes all activities that are performed to carry out the agreed terms of an order.
Definition	**Start** event: Placing an Order. The process starts when an order is placed either internally or externally. **Transformation:** Includes all activities as listed below **End** event: The customer has received and accepted the product. Furthermore he has accepted the invoice. This can either be indicated by a document or the enterprise has received the payment for the product.
Goal	The goal of order fulfilment is to fulfil all terms of an order by a customer, including the agreed quantity and quality at a certain price and to a specified date.
Remarks	For the definition of the end event the final payment is used, except for holding back money for warranty reasons.
Basic Performance Indicators	▪ Order fulfilment lead time ▪ Outgoing delivery completeness ▪ Outgoing delivery timeliness
Additional Performance Indicators	▪ Average order value ▪ Inventory cost ratio ▪ Materials cost ratio ▪ Outgoing delivery quality ▪ Backlog
Sub-processes	Operational planning, Production/Supply of products, Distribution, Invoicing
Activities	Order processing, Production planning & control, Procurement and inbound logistics, Manufacturing and assembly, Distribution and outbound logistics, Invoicing and payment.

The terms *efficiency* and *effectiveness* have been used by many authors when describing what a performance measure is intended to measure. Efficiency measures how economically the firm's resources are utilized when providing a given level of customer satisfaction while effectiveness refers to the extent to which customer requirements are met. For example, in a customer order fulfilment process, there is usually a delivery process whereby the product ordered by the customer is delivered. In this delivery process, an efficiency measure would be used to measure the throughput of this process, while an effectiveness measure would be used to see if throughput was compatible with actual requirements. Appendix to this book lists a typical selection of performance measures for business processes and competitive dimensions considered by the AMBIT model introduced in Chapter 3.

2.5 Evaluation Criteria for Performance Measurement Systems

Efforts to improve the performance of companies have been important since the
start of the Industrial era.   – Grunberg 2003

While many different performance measurement systems exist, not all are
appropriate. The headings under which a performance measurement system can be
assessed for suitability include: Framework, Business Processes, Performance
Measures and Strategy and Customers. In Section 2.6 we will examine four
contemporary performance measurement systems under these headings.

2.5.1 Framework

A framework is an important requirement for any performance measurement system.
This framework should allow:
- the movement from a high level perspective to a low level perspective,
- the identification of all business processes for a particular domain,
- the identification of performance measures (PMs) at different levels within a
  specific business process,
- the selection of processes and measures related to specific high level objectives,
- the translation of the organisational strategy and customer requirements into a
  set of performance measures.

A framework should facilitate the movement from a high level perspective to a
lower level perspective, thus allowing greater levels of detail to be exposed at
successively lower levels. This framework should support the identification of
performance measures both for any business process and at different levels within
the process itself. Furthermore, the framework should allow any performance
measures to be identified for a specific application domain (e.g. manufacturing). For
example, if the PM framework is being used to identify measures related to
manufacturing organisations, then the PM framework should allow PMs to be
identified for any part of the manufacturing organisation.

The organisational strategy and/or customer requirements should be used as the
basis for selecting a process (and its associated macro measures of performance i.e.
time, cost, quality, flexibility and the environment). The individual performance
measures themselves should also be related to either the organisational strategy or
customer requirements. Therefore, the PM framework should allow the translation
of the strategy and customer requirements of the organisation into a set of business
processes along one or more macro measures of performance.

A PM framework should provide a structured and logical approach to the
identification of performance measures while also allowing a set of performance
measures that are related to a specific objective to be identified. A PM framework
should allow the movement from a high level perspective to a lower level
perspective and provide a means for identifying performance measures at various
levels during this movement. The framework also provides a means of translating
the strategy and customer requirements into a range of performance measures.

2.5.2　Business Processes

Business processes should be broadly defined and have sufficient depth to facilitate their understanding or modelling.

2.5.3　Performance Measures (PMs)

PMs should be capable of measuring any business process found within an enterprise. This means that the PMs should measure specific activities that occur within each process. All of the performance measures used to measure a business process should be capable of being grouped under one or more macro measures of performance (e.g. Time, Cost, Quality, etc.). Before measuring a process, the macro measure(s) of performance should be identified. This would also allow the specific measurement focus for the process to be identified. For example, if the macro measures of performance for a design process are time and quality, then all of the performance measures used to measure that process should be time and quality related[4]. The performance measures used to assess the performance of a process should also be related to either the organisational strategy or the requirements of the organisation's customers. Therefore the performance measures used should have the following characteristics, namely:

- they measure specific process activities,
- they are related to specific macro performance measures, and
- they are related to the strategy of the organisation and/or the requirements of the organisation's customers.

There are two main approaches to performance measurement. The first approach involves rating the performance of specific actions and the second involves quantifying the performance of a specific action. The difference can be shown using a simple example. If the delivery performance of an organisation is being measured, a rating approach would ask the user (to rate) how he believed the delivery process performed.

The question could be phrased as follows:

> How would you rate {1 (very bad) – 7 (excellent)} the delivery performance of the organisation?

Using a quantitative approach, the delivery performance might be expressed as follows:

> Delivery Performance = 95%

In both cases, an indication of the performance of the delivery process has been obtained. Both of these approaches can also be used to identify the level of

---

[4] A set of performance measures for each of the macro business processes are given in Appendix A. For each of these business processes, the measures have been sub-divided using the macro measures of performance time, cost, quality, flexibility and the environment.

performance improvement required for specific actions or activities. However, when attempting to re-engineer a business process the rating approach to performance measurement is not very suitable. In order for a process to be re-engineered successfully, the level of improvement for specific performance measures needs to be explicitly defined. This cannot be done using the rating approach. Imagine trying to design a process where you need to improve delivery performance from a one to a seven (i.e. very bad to excellent). The only way to specify the improvement levels required for specific business processes is to use the quantitative approach and to state explicitly the level of improvement sought. This can be done by defining a target improvement level. Therefore, the performance measurement approach best suited for in this scenario will be quantitative.

In synopsis, the use of process-based performance improvement systems can result in some/all of the following [Chan *et al.* 2003, Kueng 2000]:
- Identification of potential problems in operations, so that corrective action can be taken before these problems worsen.
- Support the testing of strategies.
- Assist in directing the attention of management.
- Facilitate in resource allocation.
- Improve the communication of process objectives.

2.5.4 Strategy and Customers

All business processes selected should be related to the strategy[5] and/or the customer requirements of the organisation. Therefore, any changes made to the business processes within an organisation should reflect either the organisational strategy or meet the requirements of the organisation's customers. Davenport states that a correctly designed business process has the voice and perspective of the customer "built-in" [Davenport 1993]. In other words, we need to tackle two separate issues, namely:

1. Are the business processes selected related to the strategy and/or the customer requirements?

2. Are the performance measures used to measure this process related to the strategy and/or the customer requirements?

2.6   A Review of Performance Measurement Systems

In this section, an overview of four popular performance measurement systems used by manufacturing organisations is presented. The purpose of this assessment is to highlight their respective features so that judgment on their suitability or applicability in a specific environment can be made. We will now briefly review each of the following approaches using the four macro criteria (viz. *Framework,*

---

[5] The strategy of the organisation reflects the direction in which management believe the organisation should move.

Business Processes, Performance Measures and *Strategy & Customers*) identified in Section 2.5:

- The TOP Approach.
- The ECOGRAI Approach.
- The "Balanced Scorecard" Approach.
- The PMQ Approach.

It will be seen that while each of these performance measurement systems focuses on specific attributes, no single performance measurement system addresses all of the attributes that are required for a performance measurement system. Therefore, the eventual selection of an approach will be dictated by the type of processes one is attempting to measure.

2.6.1 The TOPP Approach

TOPP is a performance measurement system for manufacturing industry that was developed by SINTEF, Norway [Moseng and Bedrup 1993]. It has been used to measure the performance of manufacturing enterprises throughout Europe. This performance measurement system is divided into three separate parts. The first part is used to obtain an overview of the organisation being assessed while the second part is used to obtain a consensus on how the organisation operates. This consensus is obtained by surveying middle managers within the organisation. The third part is concerned with focusing on specific areas within the organisation that need improvement and attempting to define the level of improvement sought in these areas. This is done by focusing on a variety of different aspects of the manufacturing enterprise, namely: marketing, material logistics, design, technological planning, production planning & control, manufacturing/assembly, product development, top management, quality management, financial management, personnel management, maintenance, information technology, research and development, improvement processes, products, facilities, equipment, personnel and organisation. [SINTEF 1991]

For each of these areas, a series of questions are asked (see Table 2.2). The answers to these questions, on a scale of 1(very bad) to 7(very good), provide an indication as to the areas of the organisation which need improvement and the direction and level of this improvement.

Table 2.2 The TOPP questionnaire table area [SINTEF 1991].

	Status today	Realistic status in 2 years	Importance for the organisation
Question 1... (related to a specific issue)	1–2–3–4–5–6–7	1–2–3–4–5–6–7	N – M – G
Question 2...	1–2–3–4–5–6–7	1–2–3–4–5–6–7	N – M – G
...			
Question N... (related to a specific issue)	1–2–3–4–5–6–7	1–2–3–4–5–6–7	N – M – G

The questions, for example, in the material logistics improvement area could be related to:

- Supplier network with respect to product quality.
- Supplier network with respect to price.
- Supplier network respect to lead times.
- Supplier network with respect to delivery precision.
- Supplier development (active co-operation with the suppliers).
- Supplier evaluation (evaluation of new and existing suppliers).

TOPP Framework

The TOPP approach is closely coupled to a very basic framework. The framework is composed of three stages, which are used to develop a profile of the organisation. The first stage provides an overview of the organisation while the second stage is used to generate a consensus on how the organisation operates. The third stage focuses on specific areas for improvement within the organisation.

The TOPP framework, upon which the TOPP measurement approach is founded, is not suitable for business process re-engineering. The improvement areas do not encompass all specific areas of interest to industrial organisations. For example, assuming a process focus, most industrial organisations have a customer order fulfilment process yet the only TOPP improvement area that addresses this critical process is 'production planning and control'. The TOPP framework does not provide any means of identifying new improvement areas and of defining performance measures for these new areas. The user of the TOPP framework is constrained to choosing from the twenty improvement areas offered. The TOPP framework provides no means of allowing the strategy of the organisation or their customer requirements to be mapped to the specific improvement areas or to the performance measures chosen for each of these areas.

TOPP Business Processes

The TOPP performance measurement approach does not focus on business processes. Instead it focuses on twenty specific improvement areas, the majority of which are functionally oriented. The improvement areas identified by TOPP focus on specific areas of an organisation and ignore other areas. For example, again assuming a process focus, the TOPP improvement areas can be mapped to the design, manufacturing and order fulfilment processes while ignoring potential processes like material supply and co-engineering [Bradley *et al.* 1996]. Furthermore, the focus of the improvement areas is very uneven as seven of the TOPP areas can be mapped to the design process and three to the manufacturing process while one TOPP area can be mapped to the order fulfilment process.

TOPP Performance Measures

The TOPP approach uses a questionnaire to assess the performance of twenty specific improvement areas for an organisation. For each of the improvement areas, a user is asked to answer a series of questions by rating the current status of the organisation with regard to a specific aspect (e.g. inventory control or procurement competence). The TOPP user is then asked to rate the status of this aspect in two years time and to rate the overall importance of this aspect for the organisation. A sample of the TOPP approach is shown in Table 2.2. Before a process can be re-engineered, specific improvement levels for a range of performance measures need to be identified. These improvement levels must be very explicit and quantitative in nature.

The performance measures used within the TOPP approach cannot be related to either the strategy or the customer requirements of the organisation. While some of the TOPP measures may actually be related to either the strategy or customer requirements, no means of identifying this relationship is provided. Also, the TOPP measures are not process oriented. All of the performance measures used in a particular TOPP improvement area are specific to that area and all measures are deemed to be at the same level. This results in a large number of performance measures (over 800 performance measures can be found throughout the twenty improvement areas). This number of performance measures seems quite large even for measuring the current performance of an organisation or for benchmarking its performance against other organisations. Unfortunately, TOPP provides no means of filtering these measures into a smaller, critical subset of measures. Clearly, the range of measures used by TOPP are not suitable for re-engineering.

For re-engineering purposes all of the performance measures should be related to a few high level macro measures of performance. Once these macro measures have been identified, only their 'children' should be used to measure a specific process. There are no macro measures of performance defined for the TOPP improvement areas and hence all measures are considered in each area. Contrarily, by defining macro measures of performance, the number of performance measures used to measure a process can be reduced and focused.

TOPP Strategy and Customers
The TOPP approach does not account for either the strategy of the organisation or the requirements of the organisation's customers. The choice of the improvement areas offered by TOPP is not related to the organisational strategy but is instead a subjective opinion as to what the surveyed individuals (within the organisation) perceive the strategy either is or should be. Also, each of the improvement areas attains a relative importance depending on who is surveyed. For example, any marketing people will probably view marketing as being of greater importance to the firm than other areas (e.g. product development). The TOPP approach does not consider the requirements of the customer in any way, which is unfortunate, as one of the objectives of identifying the important improvement areas (and in changing them) should be to meet the requirements of the customer.

2.6.2 The TOP Approach

The TOP (Towards Optimal Performance) measurement approach was developed as part of a project funded by the Dutch Ministry for Economic Affairs [Gijben 1994]. The goals of the TOP project included the development of a single performance measure that could be used by organisations and their suppliers, to assess their logistics performance and to reduce the costs involved in the logistics process. The TOP approach is specifically concerned with the logistics process and has resulted in the identification of a performance measure that can be used by organisations to measure the delivery reliability of their suppliers. This delivery reliability measure has been proposed (by the TOP participants) as a logistics measurement standard. The TOP approach has a number of measurement points, which are used to calculate this delivery reliability measure, including:

- customer number / supplier number,
- date of order,
- supply condition,
- requested delivery date,
- agreed delivery date,
- number of products ordered.

TOP Framework
Within the TOP approach, no framework exists that can be used to logically identify a series of performance measures. No capability exists to translate either the organisational strategy or the customer requirements into a set of performance measures. The focus of the TOP approach is to develop a single measure for delivery reliability and the focus is specifically on the logistics process. The TOP approach does not consider other potential dimensions of performance (e.g. time, quality, etc.) unless they specifically relate to the delivery reliability of the logistics process.

TOP Business Processes
The TOP approach specifically focuses on the logistics process and on measuring the delivery reliability of this process. It does not consider any of the other processes that are found within manufacturing enterprises

TOP Performance Measures
The TOP approach is specifically concerned with measuring the delivery reliability of the logistics process between an organisation and its suppliers. The delivery reliability measure is given by the formula $L = L_t \times L_c$, where L_t is the timeliness of the order that has been received and L_c is the completeness of the received order. This measure alone is not sufficient for re-engineering a logistics process. This is due to the fact that it is a form of rating, as the final delivery reliability value (L) is dependent on two totally separate factors and while either of these values (L_t or L_c) may change, the overall value (L) may remain the same. The TOP approach does not identify any high level (or macro) performance measures and no consideration is given to any other dimension of performance other than delivery reliability. Also there is no linkage between the performance measure(s) used and the strategy of the organisation or the customer requirements. The measures used within the TOP approach cannot be used by any other business process.

TOP Strategy and Customers
The TOP system does not consider the strategy of the organisation or the requirements of the organisation's customers. It focuses specifically on the logistics process. Therefore it provides no means of translating the strategy or customer requirements into performance measures.

2.6.3 The ECOGRAI Approach

The ECOGRAI approach was developed by the GRAI laboratory for measuring the performance of organisations. The ECOGRAI approach allows a set of global objectives for a particular function (or organisation) to be defined. For each of these functions, a set of decision variables is identified. Management perform actions upon these decision variables which ensure that the objectives defined for the function are achieved. A set of performance indicators is identified that allow the actions that are performed upon these decision variables to be measured. The ECOGRAI approach uses GRAI grids and nets to decompose the global objectives to lower level objectives and to identify the decision variables and performance indicators required at these different levels. The ECOGRAI approach is shown in Figure 2.4.

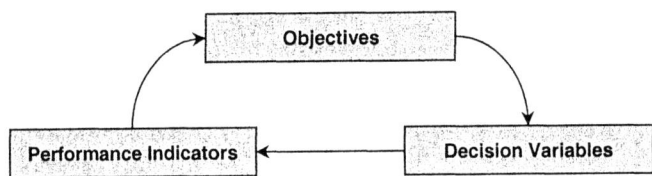

Figure 2.4 The ECOGRAI approach

The result of using the ECOGRAI approach is the development of a performance indicator system that can be used by industrial organisations. This performance indicator system can be used to assess the performance of the organisation relative to its overall objectives. Each of the performance indicators identified is described by a specification sheet which contains a variety of information about the performance indicator including indicator name, the objective that the indicator satisfies and the origin of the values used to generate the indicator. The implementation and operation of the performance indicator system is supported by an executive information system (EIS) tool.

Using the ECOGRAI approach, a set of objectives for the organisation (or functions within the organisation) can be defined at the strategic, tactical and operational level. Once the strategic objectives have been defined for a particular function, these strategic objectives can be decomposed into lower level tactical and operational sub-objectives. For each of these decisional levels (i.e. strategic, tactical and operational), it is possible to define a set of decision variables and performance indicators. However, the identification of the decision variables and performance indicators is mainly done at the operational level. The ECOGRAI approach ensures that a coherent distribution of performance indicators covering the various functions and the various decision levels exists.

ECOGRAI Framework
The ECOGRAI approach uses a framework that allows a set of global objectives to be decomposed into a series of lower level objectives which can be measured using a set of performance indicators. The ECOGRAI framework follows a top down decomposition approach and allows performance indicators to be identified for any

function in an organisation. Whilst, the framework does not allow customer requirements to be measured, the organisational strategy is measured via the high level objectives. As the decomposition approach used to obtain the performance indicators is complex and cumbersome to use, an executive information system (EIS) (which has automated the ECOGRAI approach) is required to successfully use this approach.

ECOGRAI Business Processes
The ECOGRAI approach is functionally oriented. It uses the GRAI grids and nets to break down a set of objectives for each of the functions within an organisation to lower levels. This approach also seeks to identify decision variables that can be acted upon in order that a particular function can meet its specified objectives. The ECOGRAI approach is not focused on business processes.

ECOGRAI Performance Measures
The performance indicators used in the ECOGRAI approach are based on measuring how specific decision variables change, in order that the objectives at that level are achieved. No high level macro measures are used and the measures at lower levels may not be related to any higher level macro measures that have been defined. This is because of the need to decompose the global objectives to a number of different levels with the possibility of defining performance indicators to measure how these objectives are met at specific levels. The performance indicators which are used, are generally quantitative in nature as their objective is to quantify how the organisation has performed with regard to its objectives. Also, the performance measures which are identified, are related to the strategy of the organisation which is expressed as a set of objectives. There is no means of assessing whether any of the measures are related to customer requirements.

ECOGRAI Strategy and Customers
The ECOGRAI approach provides a framework that allows a set of high level objectives to be decomposed into a set of lower level objectives. These high level objectives are defined for the various functions within an organisation and are similar to an organisational strategy, as they express the considered view of the organisation as to what it considers as being important. The requirements of the organisation's customers cannot be accounted for using the ECOGRAI approach.

2.6.4 The "Balanced Scorecard" Approach

The 'Balanced Scorecard' approach was first proposed by Kaplan & Norton [1992] and is based on the notion that a performance measurement system should provide managers with sufficient information to answer a number of questions, namely:
- How do we look at our shareholders (financial perspective)?
- What must we excel at (internal business perspective)?
- How do our customers see us (customer perspective)?
- How can we continue to improve and create value (innovation and learning perspective)?

These four separate perspectives are shown in Figure 2.5. Many commercial consultants claim to have successfully used the 'Balanced Scorecard' approach both internally and with a number of their clients.

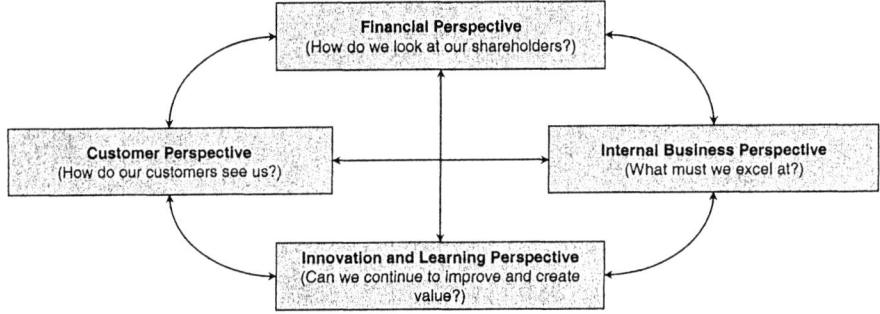

Figure 2.5 The "Balanced Scorecard" approach [Kaplan and Norton 1992]

Balanced Scorecard Framework
The 'Balanced Scorecard' approach seems, at first glance, to be well founded and broad. However, on closer examination several major shortcomings can be identified. While this framework identifies the need to consider both the strategy of the organisation (internal business perspective) and the customer requirements (customer perspective), the framework provides no guidelines or approach about how these factors should be considered. The framework also provides no approach for identifying performance measures at any level. The framework is too high level and unstructured and provides no means of focusing on lower level issues. Overall this framework is very general and is not really suitable for developing performance measures.

Balanced Scorecard Business Processes
The 'Balanced Scorecard' approach does not focus on business processes. It identifies a number of perspectives that need to be accounted for, but there is no focus (or perspective) relating to business processes. The focus could just as easily be functionally oriented. No measurement approach is defined in this framework and since the framework could be used to measure either processes or functions, then this framework does not allow process oriented performance measures to be identified.

Balanced Scorecard Performance Measures
No measurement focus is explicitly identified in this approach and no guidelines are provided as to whether macro measures should be identified or whether quantitative measures should be used. Furthermore, no approach exists for relating the performance measures to either the strategy of the organisation or the customer requirements.

Balanced Scorecard Customers and Strategy
Whilst this approach identifies the need to account for both the customer perspective and the strategy of the organisation (internal perspective), no approach for doing so

is outlined. Furthermore, no approach for identifying performance measures that are related to the organisational strategy or customer requirements is included.

2.6.5 The PMQ Approach

The Performance Measurement Questionnaire (PMQ) was developed as part of the Boston University Manufacturing Roundtable project [Dixon 1990]. The purpose of the questionnaire was to gather data about the approaches to manufacturing performance measurement used by manufacturing organisations. The questionnaire is divided into four separate parts. The first part asks a series of questions which, when answered, provide a profile of the responding organisation. In the second part, 24 areas for improvement are identified, some of which include: overhead cost reduction, manufacturing strategy, volume flexibility, computer integrated manufacturing and information systems. For each of the twenty four improvement areas, respondents are asked to rate the long term importance of each of these areas for the organisation and to assess whether the current performance measures either inhibit or support improvement in these areas.

In the third part of the PMQ approach, respondents are provided with over 39 performance improvement factors which can be used to evaluate the performance of the organisation. Some of these factors include: cost of quality, on-time delivery, number of suppliers, number of material part numbers and inventory turnover. For each of the performance improvement factors, respondents are asked to rate the importance to the organisation, of achieving excellence in this performance improvement factor and to rate the extent to which the organisation presently emphasises measurement of each of the performance improvement factors. The fourth part of the PMQ approach involves describing the primary factors that are used to evaluate performance over specific time periods (e.g. daily, weekly, monthly, etc.).

Whilst the performance measurement questionnaire tends to cover a lot of areas, there are several problems with the way in which the questionnaire is structured. These problems are as follows:
- There is no link between the strategy of the organisation, the perspective of the customer, the performance factors and the improvement areas identified in the questionnaire.
- Any of the improvement areas or performance factors can be chosen at random and may not actually reflect the actual problems within that organisation.
- Given the range of managers surveyed, the questionnaire may produce certain bias, as managers may want to back their own areas of responsibility.

The performance measurement questionnaire, whilst reasonably complete, provides no means of identifying the processes within the organisation or of identifying specific performance measures that are related to the strategy of the organisation and its customers' viewpoints.

PMQ Framework
The performance measurement questionnaire (PMQ) uses a basic framework consisting of four parts. The first part of the framework is used to build a profile of the manufacturing organisation while the second part is used to identify specific areas in which an organisation is trying to improve its performance. The third section is used to identify specific improvement factors that the organisation is using in an attempt to improve its overall performance, while the fourth part is used to obtain a profile of how performance is evaluated over specific time periods. The PMQ framework does not provide a top down focus and all of the performance improvement areas are at the same level. Furthermore, the framework does not allow the translation of either the strategy of the enterprise or the requirements of the organisation's customers into a set of performance measures that could be used to assess the performance of the enterprise. This framework does not address all possible improvement areas within manufacturing enterprises. The PMQ user is constrained to choosing from one of the proffered improvement areas. It is not possible to identify other improvement areas using this framework.

PMQ Business Processes
The PMQ approach is not process focused. This approach expects the user to identify the long run importance of 24 improvement areas in the manufacturing enterprise. None of these areas are process oriented and the PMQ approach does not allow any other improvement areas to be identified. Also, assuming a process focus, a large majority of these improvement areas identified seem capable of being grouped into design and manufacturing oriented processes.

PMQ Measures of Performance
PMQ uses a rating approach to performance measurement. For measures like "Manufacturing lead times" and "On-Time Deliveries" the user is asked to RATE the importance of this measurement to the organisation and then to RATE the emphasis the organisation places on this measure. A sample of the PMQ measurement approach is shown in Table 2.3.

Table 2.3 A sample of the PMQ measurement approach [Dixon 1990]

Importance of Performance Factor None (1) ➔ Great (7)	Performance Factors	Organisation's Emphasis on Measurement None (1) ➔ Great (7)
1 – 2 – 3 – 4 – 5 – 6 – 7	Inventory Turnover	1 – 2 – 3 – 4 – 5 – 6 – 7
1 – 2 – 3 – 4 – 5 – 6 – 7	Cost of Quality	1 – 2 – 3 – 4 – 5 – 6 – 7
1 – 2 – 3 – 4 – 5 – 6 – 7	Vendor Lead Times	1 – 2 – 3 – 4 – 5 – 6 – 7
1 – 2 – 3 – 4 – 5 – 6 – 7	Sales Forecast Accuracy	1 – 2 – 3 – 4 – 5 – 6 – 7
1 – 2 – 3 – 4 – 5 – 6 – 7	On Time Deliveries	1 – 2 – 3 – 4 – 5 – 6 – 7
1 – 2 – 3 – 4 – 5 – 6 – 7	Number of Suppliers	1 – 2 – 3 – 4 – 5 – 6 – 7
1 – 2 – 3 – 4 – 5 – 6 – 7	Unit Material Costs	1 – 2 – 3 – 4 – 5 – 6 – 7

As earlier outlined, the performance measures used in the PMQ approach are not related to either the strategy of the organisation or to the requirements of the organisation's customers. The measures are not capable of measuring specific process activities and no means of identifying new performance measures is provided. All of the performance measures are deemed to be on the same level and no capacity for identifying high level (or macro) measures of performance exists.

PMQ Strategy and Customers
The PMQ approach provides no facility for identifying improvement areas that are specifically related to either the strategy of the organisation or the requirements of the organisation's customers.

2.7 Developing a Performance Measurement Strategy

All performance measures should relate to the performance of a process (or activity) and how well this process (or activity) is performed. However, in reality this is often not the case, as some organisations tend to use performance measures like 'scrap rate', 'overtime cost', 'product cost', etc. as the basis of their decision making. Such measures represent a throwback to the traditional performance measures and to an era when there was no real linkage between manufacturing and business strategy (i.e. manufacturing collected their own performance measures and made decisions based on those measures, while business made decisions based on other sets of measures). The design of any performance measurement system should reflect the basic operating assumptions of the organisation it supports. If the organisation changes so also should the measurement system. If not, then the performance measurement system could at best be ineffective and at worst counterproductive [Neely 2003, Meyer 1994]. As many organisations move from a functional, control oriented system to a team based process oriented system, traditional performance measurement systems not only fail to support the new teams but they also undermine them.

Performance measures are best developed co-operatively by those responsible for the system being measured. All affected members of the organisation should participate in order to understand the desired outcomes and the level of expectations. Involvement should be at all stages i.e. developing measures, monitoring performance on a continued basis, self-assessment and evaluation. Continuous interaction prevents surprises and the focus is shifted to performance as opposed to compliance. Therefore, when developing a performance measurement strategy it is important to consider the following:

- **Measure only what is important**. Do not measure too much, measure things that impact the bottom line, such as, customer satisfaction.
- **The measurement system as a whole must be implemented**, not just one or two easy or attractive measurements picked out.
- **Focus on customer needs**. Ask your customers what they think you should measure.

- **Co-ordinate the performance measurement system** with other impacted organisations where appropriate.
- **Secure employee buy-in** by giving them a sense of ownership. This will lead to improvements in the quality of the measurement system.
- **Make sure that the measurements are stable** i.e. not changed to due to internal political pressures or flavour of the month ideas.

Motivational theory suggests that individuals respond positively to stimuli that reward achievement and performance. Employee recognition and compensation are key elements in aligning the interests of employees to that of the organisation. For example, if organisations wish to encourage certain activities, such as teamwork, they must reward team members for their activities.

Performance measurement tools (and the corresponding reward systems) are useful for aligning goals with an organisation's objectives. The purpose of the rewards system is to align the goals of the employee with the goals of the organisation. The organisation's reward system provides motivation and an incentive for meeting strategic objectives. In other words, rewards deal with motivation, which determines behaviour, which, in turn, affects performance. For an organisation to be effective, all the policies must be aligned, interacting harmoniously with each other. An alignment of the reward system with the performance measurement system will communicate a clear consistent message to the organisation's employees.

The development of performance measures is an evolutionary process. As strategic plans are revised, and performance measures are added, dropped or revised, a lag in conformity is expected. This should be recognised as inevitable and the sign of a healthy, growing, continuously improving system.

2.8   Performance Measurement in Extended Manufacturing

Performance measures discussed up to now are valid if one considers their implementation in a single enterprise. However few modern manufacturing organisations exist in isolation. They tend to collaborate very closely with their customers and suppliers in a so-called extended enterprise environment. Therefore, all benchmarking and performance assessment issues an organisation is considering should take into account this extended environment[6].

Furthermore, to tap the true benefits from a performance measurement system, it should be closely linked to the rest of the organisation's information and decision support systems and preferably be designed to enable automatic collection of the performance data. Many organisations use some form of Enterprise Resource Planning (ERP) software to manage their day-to-day operations. Therefore, it makes sense if the PM system resides side-by-side, as perhaps as an application layer on the ERP software. If this architecture were followed, a framework to exploit real-

---

[6] See chapter 3 (Figure 3.8) for a typical extended enterprise model and a discussion on the notion of extended enterprise.

time, operational performance measurement combined with performance benchmarking can be developed for daily management and work improvement.

An ERP system is an integrated Enterprise Resource Planning systems that functions as central nerve system of an enterprise, and is the basis for most of the decision-making. By tapping into the data available in these systems and calculating standard performance indicators based on it, performance measurement can be greatly facilitated.

While considering an automated and linked performance measurement system, one needs to consider two attributes:

■ A *performance measure* (PM) is a basic figure that is collected from the enterprise. This figure is not used to represent performance as such. Two typical examples are "number of customer order lines delivered on time" and "total number of customer order lines delivered".

■ A *performance indicator* (PI) is frequently created by performing calculations involving two or more performance measures. The indicators are the entities that are monitored, analysed, and compared when managing performance. Based on the two performance measures mentioned above, the corresponding performance indicator "outgoing delivery timeliness" is calculated by finding the ratio between the two.

Bearing these issues in mind, one needs a generalised performance measurement system, which can be tuned by an individual enterprise to suit its internal and external (supply chain and business networks) requirements. One solution is to develop the performance measurement system along three dimensions, as depicted in Figure 2.6.

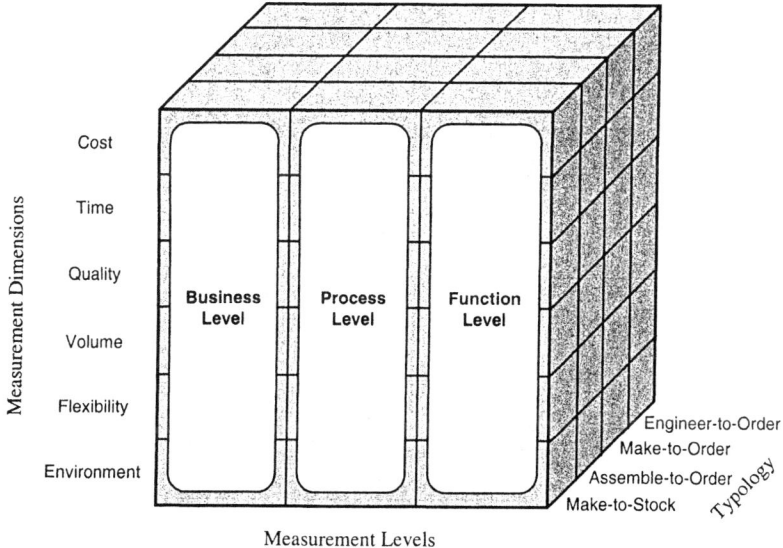

Figure 2.6 Performance measurement cube

This approach, developed by ENAPS consortium [ENAPS 1999] satisfies the criteria set (listed in Sections 2.4 and 2.5), can truly allow networked organisations to manage their performance and identify necessary improvements.

The performance measurement cube[7] (Figure 2.6) outlines three dimensions that should be considered when measuring performance. These dimensions are enterprise typology[8] (ranging from engineer-to-order through make-to-order to assemble-to-order to make-to-stock), measurement levels (business, process, and function level), and measurement dimensions (cost, time, quality, volume, flexibility, and environment). At each location in the cube, the best-suited performance measure and indicator will not only be process specific but will also be uniquely different for the industry sector and the enterprise involved.

The performance measurement framework illustrated in Figure 2.6 needs to be supported by a detailed business process framework that specifies a set of generic business processes (Figure 2.7). These processes are divided into two types of business process; the core value-adding processes of the enterprise, and secondary processes, that support the core business processes.

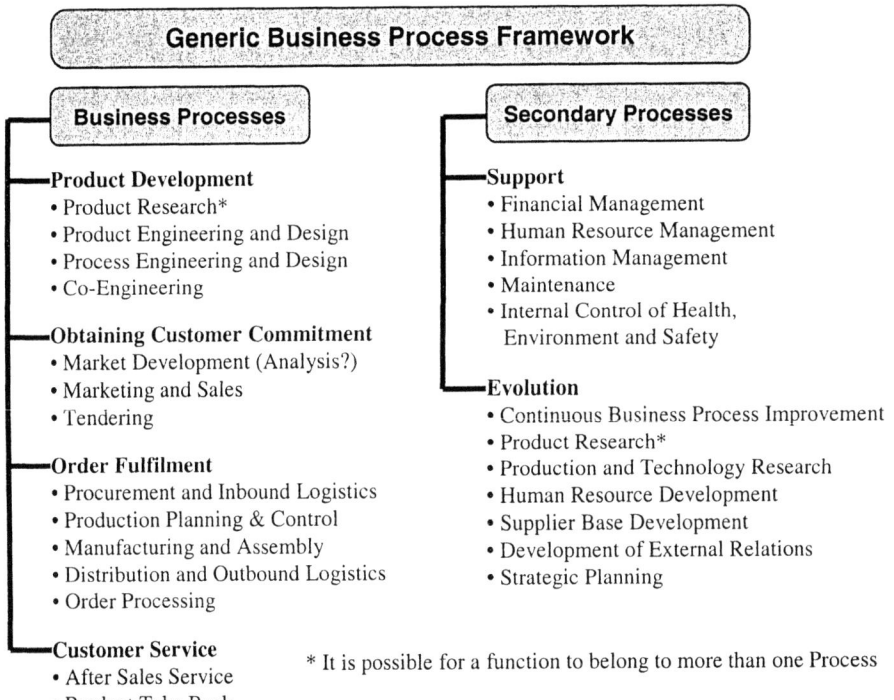

Figure 2.7 Generic business process framework [ENAPS 1999]

---

[7] We will use a modified version of this approach in AMBIT. See Chapter 4.
[8] We will discuss manufacturing typology in detail in Chapter 4 (Section 4.3.3.).

An automated performance measurement system should be flexible and be able to evolve as an enterprise changes or reconfigures itself within a network, to utilise the new opportunities the systems present. This will most likely take on the form of an overall model combined with loose process models that depict what the automated performance measurement can do and how the user should go about to exploit these options. Thus, the result may very well be some form of flow diagrams that present the user with some key questions on what is to be achieved and then explain how this can be done. Explanation to focus on should include:

- How to determine which performance measures and indicators (PMs/PIs) to make use of, and how to interpret these?
- How to use the performance measurement for decision support in daily operational management within one single enterprise?
- How to conduct diagnostic self-assessment using the performance measurement?
- How to conduct performance benchmarking in the enterprise?
- How to exploit the performance measurement for process improvement work in one enterprise?
- How to use the performance measurement for decision support in managing the extended enterprise?
- How to assess the performance of the business network based on performance data from the performance measurement?
- How to benchmark the performance of the extended enterprise?
- How to facilitate business network improvement based on decision support offered by the performance measurement?

The potential benefits of a generalised performance measurement system that is linked to the organisation's other business processes as well as its partners can be very great indeed and include:

- The ability for an organisation to customise the performance measures and indicators.
- Access to real-time operational performance data, i.e., collected and presented within a very short time span after the events and actions that led to the performance level took place. These performance data could be collected automatically and continuously presented to the relevant decision-makers.
- Opportunities for being presented with the same type of performance data for one or more business networks that the enterprise is part of.
- The possibility to continuously benchmark performance with different sets of enterprises, e.g., within different countries, sectors, of a certain size, etc.
- The ability to define specific monitoring programmes to keep track on improvement efforts put in place.

An automated performance measurement system that is layered on to an ERP system will allow very rapid transfusion of performance data, enabling its use in operational decision support, both within a single enterprise and across networked organisations. A generalised performance measurement system should be capable of isolating industry specific measures and indicators, than can be exploited for benchmarking within industries. It should also be able to take full advantage of the available information and communication technologies to enables benchmarking within or across geographical regions.

2.9   Conclusion

Performance measurement facilitates an improved understanding of an organisation's objectives by all employees. Individuals get an opportunity to develop a broadened perspective of the organisation's functions, rather than a more limited perspective of their own immediate span of control. However, measuring the wrong things in the wrong area or at the wrong level in an organisation can prompt an inappropriate response and affect the ability to achieve organisation objectives.

Substantial benefits can be realised by organisations implementing performance measurement programmes. The following outlines some of the benefits of performance measures:

- They help organisations understand their processes.
- They identify whether an organisation is meeting customer requirements.
- They identify whether suppliers are meeting an organisation's requirements.
- They ensure decisions are based on fact, as opposed to emotion.
- They identify where improvements need to be made.
- They verify the effectiveness of corrective actions.
- They enable continuous access to objective data to support an organisation's claims of quality.
- They receive early warning of problems or conditions that could lead to serious error.
- They demonstrate accountability to all stakeholders.
- They allow organisations to benchmark their performance with other organisations.

Performance measurement is the trigger for performance improvement and the trend is towards the use of contemporary performance measures, which emphasise a range of criteria, unlike earlier approaches which were driven primarily by considerations of cost. Enterprises have found that these contemporary measures, emphasising as they do measures of time, quality, flexibility etc. as well as cost provide the information required to compete in the current global economy. A range of contemporary performance measurement approaches has been developed and a cross-section of these approaches was outlined in this chapter.

Chapter 3

Overview of the AMBIT Approach

Keywords: Manufacturing strategy, customer order fulfilment, design co-ordination, co-engineering, extended enterprise, virtual enterprise, extended products, value chain, product cycle view, Porter's competitive forces, investments, world class manufacturing, formalisation, performance measurement and the AMBIT approach.

Chapter Objectives: The purpose of this chapter is to stress the importance of the manufacturing strategy in the development of a competitive edge and to document the reasons for developing the AMBIT approach. Having read this chapter, the reader should become familiar with:

- Porter's five competitive forces.
- The importance of the manufacturing strategy. Performance of the manufacturing function influences the strategic direction of a manufacturing organisation. Therefore, it is vital that there be compatibility between the manufacturing strategy and the business strategy.
- World class manufacturing.
- The extended enterprise and its relation to Porter's model, the value chain, CIM and the total product cycle view.
- The concept of extended products.
- The reasons for the development of the AMBIT approach.
- How the AMBIT approach has been developed to fulfil the support formalisation required by managers.
- The basic outline of the AMBIT approach.

We have also included in this chapter a discussion on World Class Manufacturing, the Extended Enterprise, the Virtual Enterprise and Extended Products because such developments provide a context for decision making in manufacturing systems. In our experience many manufacturing systems operate within or as part of extended enterprises, that is they have well developed partnerships with suppliers, distributors and customers and their business processes are tightly coupled to those of their partners. Frequently, these connected business processes are mediated by advanced information and communications technology systems supporting e-Commerce applications software. One of the most interesting examples of this trend is the relationship between the contract electronic manufacturers (CEMs) and their large customer OEMs (Original Equipment Manufacturers).

3.1   Introduction

> In today's turbulent competitive environment, a company more than ever needs a strategy that specifies the kind of competitive advantage that it is seeking in the marketplace and articulates how that advantage is to be achieved.
> – Hayes *et al.* 1994

Today's organisations are moving away from static processes and strategies towards dynamic, market-driven frameworks. Clearly, there is no one framework or model which can be universally applied and which can then act as a guarantee for successful competition. There is no one benchmark or target which, when attained will ensure constant, sustainable success for the organisation. Consequently, management need to think in terms of a continuum of continuous improvement within all aspects of the organisation. Furthermore, with the increasing emphasis on information technology tools and methodologies as competitive weapons, technology investment decisions are critical. Appropriate technology decisions can support the organisational pursuit of manufacturing excellence and market success, whilst, incorrect technology investment decisions can result in large liabilities. As a result, technology investment is becoming a crucial part of the organisation's competitive arsenal.

A decision to invest in a technology and/or programme will have many far-reaching organisational ramifications, in terms of employees, skills and manufacturing systems design and operation. Thus, such a decision is supported if:
- The potential strategic impacts of this decision can be quantified.
- The investment decision is in harmony with the organisation's strategic goals.

The AMBIT (Advanced Manufacturing Business ImplemenTation) manufacturing model and the associated manufacturing business processes are based on the Extended Enterprise, Virtual Enterprise and Extended Products concepts of manufacturing systems. This model also takes into account the issues raised by the world class manufacturing (WCM) model. Hence, it is appropriate that these topics be briefly outlined before the AMBIT manufacturing model is presented.

In this chapter, the importance of the manufacturing strategy and the need for compatibility between the manufacturing strategy and the business strategy is illustrated. Furthermore, the reasons for the development of the AMBIT approach are presented and a brief overview is given of its structure.

3.2 The Importance of Manufacturing Strategy

> It is simply wrong to say that traditional manufacturing has no future ... intelligent use of new technology can lead to substantial competitive advantage for individual manufacturers. By the same token companies that fail to recognise the opportunities for new processes and better products will lose market share...Success in this new environment will require a sustained effort by those who work, manage and invest in manufacturing.   – DTI 2002

A strategy refers to a 'business game plan', whilst strategic planning involves devising necessary or suitable changes to this game plan. Strategic management entails the implementation of such changes.

Porter states that competition in industry is driven by five basic forces (Figure 3.1). "The collective strength of these forces determines the ultimate profit potential of an industry" [Porter 1995].

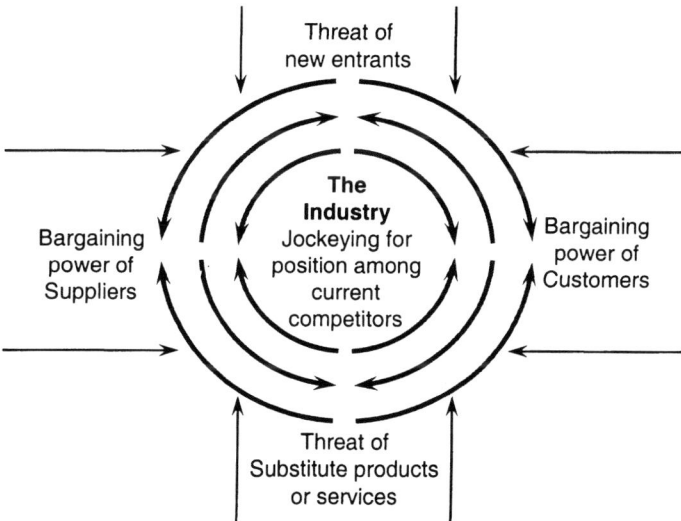

Figure 3.1 Forces governing competition in an industry [Porter 1995]

The purpose of strategy formulation is to better equip the organisation in its response to these competitive forces. Influencing this response is the awareness that (manufacturing) organisational success is not solely dependent on product design, marketing ingenuity and financial strength. Overall manufacturing capability is also extremely important. Many manufacturing organisations over the years have systematically neglected their manufacturing function, as organisational success was traditionally regarded to be mostly influenced by marketing. Manufacturing was traditionally viewed as a routine function having little or no input to the process of organisational strategy formulation.

However, in the late 1960s, Skinner identified the contribution of manufacturing to the overall success of the manufacturing organisation. He also pointed out the lack of managerial awareness concerning the strategic implications of routine manufacturing decisions; decisions which greatly impinged on the corporate success of the organisation. Resulting from managers' limited understanding of the importance of manufacturing to strategic planning, decisions of a manufacturing nature were generally delegated to those in manufacturing, with little or no attempt to verify if a link existed between such decisions and the corporate strategy. Manufacturing decisions were initiated, specified and introduced by engineers, who had insufficient information about the business context within which they operated. This neglect of the linkages between manufacturing and business strategy persists to some extent today, for despite the importance of the manufacturing strategy, there is still little input from manufacturing into corporate strategy [Mellor *et al.* 2002].

Thus, manufacturing is no longer considered solely to be the transformation of raw materials into a finished end product. Current economic conditions and market forces mean that manufacturing involves considering "the transformation of information (customer needs, design data, production data, etc.), the transformation of geography (logistics) and the transformation of availability (inventory control and lead times)" [McCarthy *et al.* 2003]. Mohd [2002] states that "although the service sector plays a major role in this ICT-era, manufacturing still plays a dominant part in creating value added in most countries. Its contribution to the aggregate economy is still significant to emulate growth". The importance of the manufacturing strategy to the business strategy and ultimately to the corporate strategy is increasingly recognised by today's manufacturing organisations (Figure 3.2).

Figure 3.2 The importance of manufacturing strategy

When we talk about manufacturing strategy, we refer to decisions relating to the management of the organisation's manufacturing operations and the deployment of the organisation's manufacturing resources; and the assessment of these decisions on the organisation's long-term objectives. Corporate strategy, meanwhile deals with the long-term management of an organisation. This includes its manufacturing operations. "Corporate strategy has been the subject of considerable academic study for more than 30 years, encompassing the consideration of both content (i.e. what the decisions and actions are) and process (i.e. how those decisions and actions come about). Manufacturing strategy, by contrast, has received much less attention, particularly with regard to process" [Barnes 2002].

In promoting a successful organisational response to today's competitive environment, management need to develop manufacturing strategies compatible with their overall business and ultimately corporate strategies. The manufacturing function should be regarded as a strategic resource, a source of strength by itself as well as a means for enhancing the contribution of other functions. Hayes and Wheelwright state that the success of the manufacturing function is not based solely on efficiency and productivity, but also on the coherence of the manufacturing strategy with its associated business strategy [Hayes *et al.* 1994].

Given that manufacturing strategy is concerned with the management of the manufacturing operations and the deployment of the manufacturing resources in order to provide competitive advantage, the introduction of new manufacturing technologies (e.g. new manufacturing processes, machines, interconnect technology etc.) and new manufacturing programmes (such as lean production, concurrent engineering, etc.) will influence the manufacturing strategy and ultimately the strategic performance of that organisation (Figure 3.3). Since technology and improvement programmes are normally implemented for the purpose of enabling the organisation to better serve its customers; current and future technology and programme trends must be regarded as important strategic considerations.

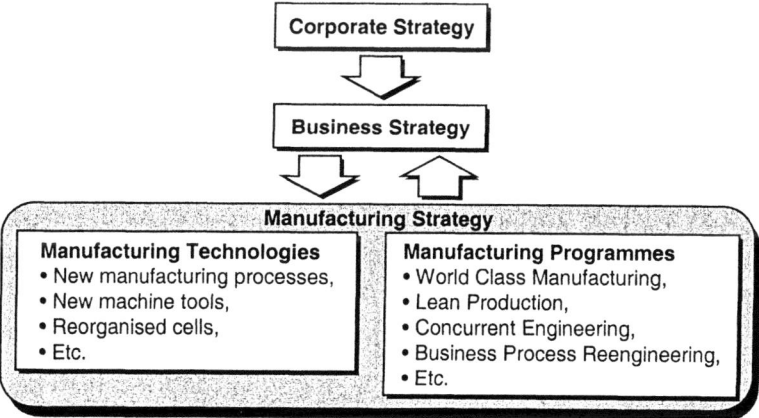

Figure 3.3 Impact of new manufacturing strategies on business strategy

3.3 World Class Manufacturing

What does it mean to be a world class competitor? It means being successful in your chosen market against any competition – regardless of size, country of origin or resources. It means matching or exceeding any competitor on quality, lead-time, flexibility, cost/price, customer service and innovation.
– MAPICS 2002

In many ways, the ideas of World Class Manufacturing [see for example Schonberger 1987, Jagdev *et al.* 1998] were developed from the experience of JIT implementations in factories in the United States of America. Issues of continuous improvement, training and cross training of personnel, integration of product design and process design to facilitate efficient manufacturing were also emphasised. Hayes, Wheelwright and Clark [Hays *et al.* 1988] for example identified the key characteristics of a world class manufacturing plant as follows (this view is also supported by MAPICS, 2002):

- **Becoming the best competitor**. Being better than almost every other organisation in the industrial sector in at least one aspect of manufacturing.

- **Growing more rapidly and being more profitable than competitors**. World class organisations are able to measure their superior performance by observing how their products are accepted in the marketplace.

- **Hiring and retaining the best people**. Having operators and managers who are so skilled and effective that other organisation are continually seeking to attract them away from the organisation.

- **Developing engineering staff**. Being so expert in the design and manufacture of production equipment that equipment suppliers are continually seeking advice about possible modifications to their equipment, suggestions for new equipment, and agreement to be a test site for one of their pilot models.

- **Being able to respond quickly** and decisively to changing economic conditions and market forces. Being more nimble and flexible than competitors in responding to market shifts and/or pricing changes and in getting new products out into the market faster than they can.

- **Adopting a product and process engineering approach that maximises the performance of both**. Intertwining the design of a new product so closely with the design of the required manufacturing process that when competitors reverse engineer the product they find that they cannot produce a comparable one in their own factories without major retooling and redesign expenses.

- **Continually improving facilities, support systems and skills** that were considered to be near optimal or state of the art when first introduced, so that they increasingly surpass their initial capabilities.

Hayes *et al.* went on to say that the emphasis on continuous improvement is the ultimate test of a world class organisation. This interest in continuous improvement later led to the notion of a Learning Organisation. In recent years these ideas have been further developed, for example, the Centre for Integrated Manufacturing Studies at Rochester Institute of Technology [Boykin 1997] has extended this thinking to define World Class Customers, Suppliers and Supply Chains. They define the salient features of each as follows:

World Class Customer (WCC):
- A WCC is a solutions developer.
- A WCC's products are conscientiously built and have zero defects.
- A WCC has effective relationships with its entire supply chain.
- A WCC is always learning and adapting.
- A WCC profits from its chain-wide co-operation and strategic vision.

World Class Supplier (WCS):
- A WCS produces solutions for its customers.
- A WCS assists in product development.
- A WCS delivers error free products.
- A WCS has a production system which delivers products on time, and knowledge which assists customers in reducing time-to-market.
- A WCS creates and sustains relationships with all members of the Supply Chain that achieve superior results.
- A WCS is an organisation that learns and adapts rapidly to respond to a world of rapid change.
- A WCS attains a return on investment which contributes to the success of all members of the supply chain.

World Class Supply Chain (WCSC):
- A WCSC develops relationships which deliver results that produce solutions for its customers
- A WCSC develops products and services that meet or exceed customer requirements.
- A WCSC defines, achieves and improves quality and safety throughout the organisation.
- A WCSC's production system delivers products on time.
- A WCSC focuses on the ability to achieve superior results by working together to satisfy customers.
- A WCSC achieves superior results through its collaborative work.
- A WCSC acts as a single organism that learns and adapts rapidly in response to a world of rapid change.
- A WCSC applies new ideas that positively affect cost, system performance, product, processes, and interfaces.
- A WCSC earns a return on investment which contributes to the success of all members of the supply chain.
- A WCSC achieves long-term business success.

3.4 The Extended Enterprise Framework

The extended enterprise can be regarded as a kind of 'enterprise' which is represented by all the organisations or parts of organisations, customers, suppliers and sub-contractors, engaged collaboratively in the design, development, production and delivery of a product to the end user [ELANCE 2002, Gott 1996, Browne 1996]. Key suppliers become almost a part of the principal company and its information infrastructures, with frequent exchange of status information.

The extended enterprise is a term used to reflect the high level of interdependence that exists between organisations as they conduct business. This is echoed succinctly in the definition of extended enterprise given in Jagdev *et al.* [Jagdev, 1998] as: "the formation of closer coordination in the design, development, costing and the coordination of the respective manufacturing schedules of cooperating independent manufacturing enterprises and related suppliers".

The keyword in this definition is the 'coordination of the respective manufacturing schedules'. This coordination of respective schedules (which includes not only the production schedules but also the dispatch, transportation/delivery and receipt notifications) is supposed to be performed seamlessly through the use of ICT technologies. Only then one can truly realise the integration of respective IT infrastructures, which is a necessary condition for the formation of extended enterprise.

The evolution of the enterprise collaboration and the sharing of business information (sharing of timely information to coordinate the decision-making is the foundation of the extended enterprise) is illustrated in Figure 3.4.

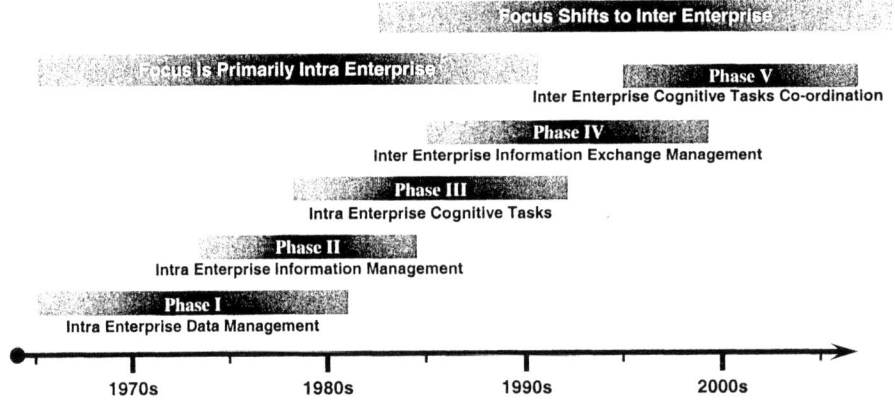

Figure 3.4 Evolution of the extended enterprise

The traditional organisational model is characterised as a single organisation composed of multiple departments (i.e. sales, engineering, purchasing, accounting, manufacturing, etc.). Meanwhile, the extended enterprise model is realised when organisations or parts of organisations, customers, suppliers and sub-contractors, are engaged collectively in the design, development, production and delivery of a product to the end user [ELANCE 2002, Gott 1996]. The extended enterprise thus includes the relationships that an enterprise has with its customers, suppliers, business partners, even erstwhile competitors and so on. It is responsible for the whole product life cycle, from material procurement and supply management, to product production, manufacturing, product distribution, customer service, recycling and disposal of end-of-life products. The extended enterprise model of manufacturing systems basically views the manufacturing system as one element of a value chain, which delivers service to a customer. It includes not only direct business partners, but also non-business partners, such as government agencies, regulation offices, academic organisations and many more. The increased availability of communications and information technology tools and the emergence of global telecomputing infrastructures support the extended enterprise model. As a result, it is important to have as much compatibility among IT systems as possible. Indeed, the extended enterprise can be considered as a driver of and a consequence of advanced information and telecommunication technology. The implementation of business-to-business (B2B) eCommerce systems is fully realised in extended enterprises.

Extended enterprises are evolutionary in nature. For example, let use take two organisations which have known each other and conducted business for some time. During this period of collaboration, a sufficient level of trust has developed to automate the sharing of day-to-day operational data. This sharing was preceded by realising the importance of each organisation to the other in its business plans. Therefore, each of the organisations will be prepared, if necessary, to invest in the modern ICT tools for the effortless sharing of information. The fact that they are willing to invest in their collaboration will imply that they are committed to a long-term relationship. It is this seamless exchange of relevant operational information on top of an existing long term (and successful) relationship that distinguishes the extended enterprise from other forms of long-term collaboration such as a supply chain relationship. It should also be noted that ICT are enabler technologies and a necessary (though not sufficient) condition for an extended enterprise to exist. It is the integration of the respective information and decision systems and the respective production processes that link them closely enough (within the agreed bounds) to be analogous to the behaviour of a single enterprise operating within four-walls.

The principle of the extended enterprise system is that its suppliers and customers are not separate entities, they are all part of the "larger us" [Browne *et al.* 1999]. Thus, whilst the supply chains and logistics chains are part of the extended enterprise, its success is greatly influenced by the speed and efficiency with which information can be exchanged and managed among the different business partners. As an example, collaborative engineering and production requires efficient electronic management of engineering and production information.

The baseline for an extended enterprise is always two or more willing enterprises which have chosen to concentrate on their core-competencies and wish to extend their activities into other enterprises to increase their competitiveness by achieving cost-, time- or quality-related advantages regarding their respective offerings. Extending the activities implies that an enterprise is enhancing its existing capabilities or adding additional facilities, by outsourcing, which has not been at its disposal so far. Outsourcing encourages both the manufacturer and its suppliers' competitive ability and enhances their mutual dependency. Figure 3.5 pictorially represents the major characteristics of the extended enterprise.

Whilst the extended enterprise clearly defines an environment where business partners work together for long-term business purposes based on mutual responsibility and loyalty, its goals include:
- The promotion of decentralised decision making and team based practises.
- The sharing of knowledge with suppliers.
- The improvement of the suppliers' performance.

- Increased product, production process and service innovation.
- The elimination of waste in the production process.
- The improvement of employee technical, analytical, social and creative competencies.

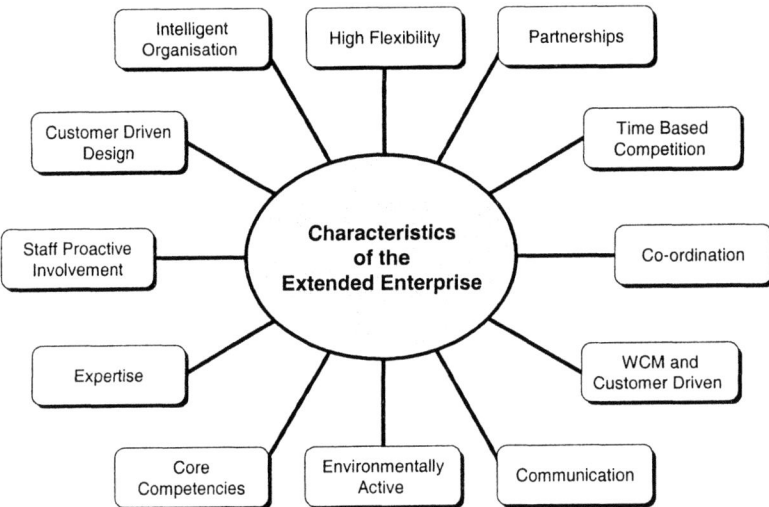

Figure 3.5 Characteristics of the extended enterprise

This environment is realised using advanced ICT (Information and Communication Technologies) systems, including the Internet and emerging B2B and indeed B2C (business-to consumer) systems.

Taken from this perspective, we can identify the following as major characteristics of the extended enterprise [Jagdev *et al.* 2001]:

- The partners in the extended enterprises are willing to form long-term relationships and treat each other as business partners. Each partner understands and accepts other's requirements and priorities.
- Within the scope of collaboration, partners share a vision and work towards a shared goal.
- The decisions are jointly arrived at by making best use of the competencies among the partners.
- The primary mode of communication and sharing of information between the collaborating enterprises will always be through telecomputing. It is, therefore, important to have available advanced ICT tools to support the extended enterprise [Bort 2003, Thommessen, 1996].
- The efficiency of the extended enterprise is greatly determined by the speed and efficiency with which information can be exchanged and managed among business partners. Efficient collaborative engineering, production and logistics require effective electronic management of engineering and production information. Thus it is important that the participating enterprises have sufficiently sophisticated IT and decision support tools and mechanisms to make the integration possible. It is also important to have the maximum degree

of compatibility among partners' IT systems. "Intermingled ownership no doubt causes problems. For instance, how can data be coded so everyone's systems can understand it easily? How can a person's identity and role be determined when location is no longer a factor? How can reliability be increased and finger-pointing minimized when a system failure occurs?" [Bort 2003].

- Day-to-day communications between the respective IT systems of two enterprises will always be real-time and on-line and without human intervention. For example, if there are production schedule perturbations in the 'customer' enterprise, these changes will (and should) be automatically communicated to the 'supplier' enterprise IT systems. Thus, triggering the processes necessary for updating of production schedules at the 'supplier' end.

- Extended enterprise can occur between any two enterprises across the value chain of any product or service. Enterprises across the whole value chain can be involved in the extended enterprise. However, the concept of the whole value chain is neither necessary nor a relevant condition to the formation of extended enterprises. If this were the case, then we would undoubtedly enter onto a very slippery slope of defining the boundaries of the value chain. For example, why restrict the starting point of the value chain to raw materials; why not extend it to their mining? Furthermore, 'raw materials' for one value chain could very well be a 'finished product' for another value chain, e.g. ball bearings, electric motors, transformers/power supply units, etc.

- Technology permitting, the extended enterprise can take the form of a complex enterprise network where each enterprise can be seen as a node.

- The relationship between a set of nodes in an extended enterprise can be hierarchical or non-hierarchical.

3.4.1 Extended Enterprise and CIM

Traditional manufacturing models concentrated on integration inside the four walls of the manufacturing plant. This was termed Computer Integrated Manufacturing (CIM). As manufacturing and distribution were generally viewed as separate entities, decoupled by warehouses and storage locations, integration between manufacturing and distribution was seldom realised (Figure 3.6).

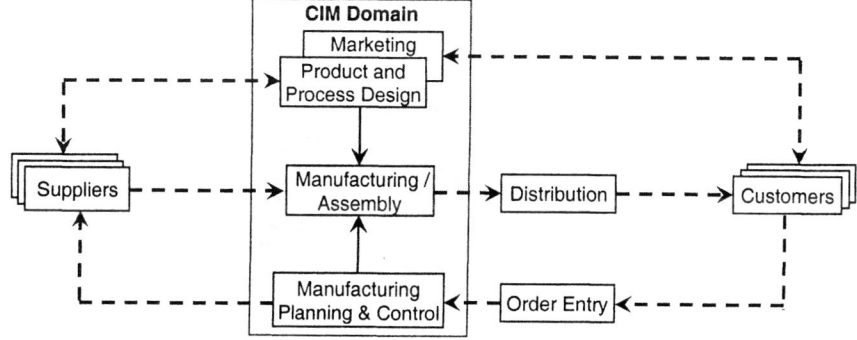

Figure 3.6 Traditional manufacturing models

With the need to support product customisation and realise shorter product delivery lead times, the focus has changed to a customer-order driven approach which integrates the manufacturing and distribution functions (Figure 3.7).

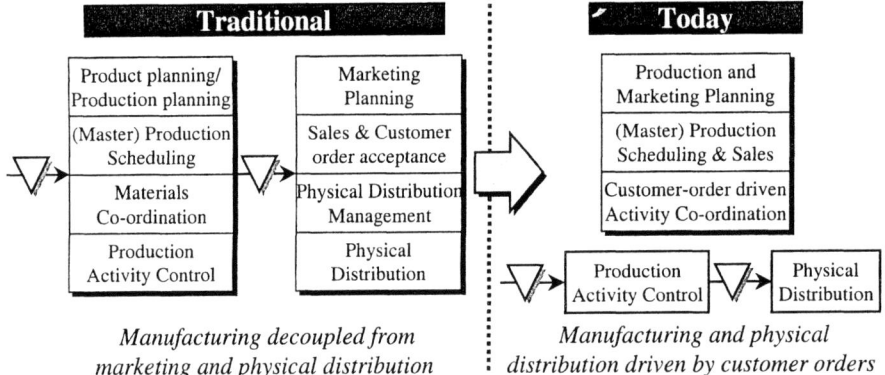

Manufacturing decoupled from marketing and physical distribution   *Manufacturing and physical distribution driven by customer orders*

Figure 3.7 The changing emphasis on the integration of manufacturing and distribution

Today's value chain is complex. It does not solely comprise one manufacturing plant, but many. Thus, the manufacturing function must look beyond "the four walls of the manufacturing plant" and across the entire value chain, i.e. to the extended enterprise model (Figure 3.8). Indeed, today's generation of ERP (Enterprise Resource Planning) and PDM (Product Data Management) systems are clearly focused on the supply- and value-chains, unlike the earlier generation of MRP (Materials Requirements Planning) and MRPII (Manufacturing Resource Planning), which were focused on the activities taking place inside the manufacturing plant.

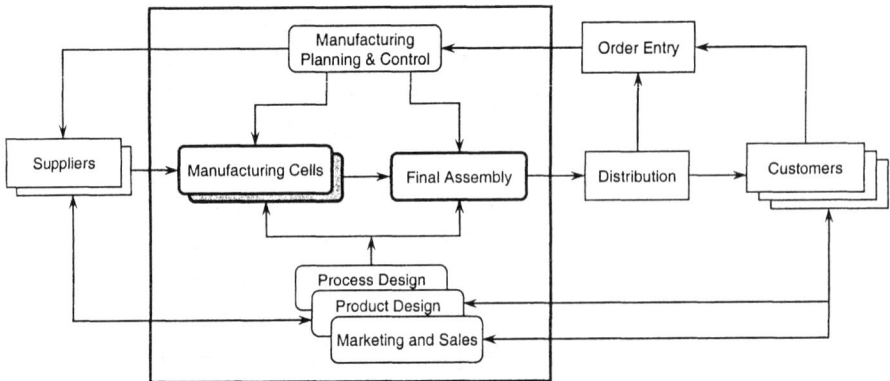

Figure 3.8 CIM and the extended enterprise

3.4.2 Extended Enterprise and the Value Chain

Any organisation can be described using the concept of a value chain [Blackhouse *et al.* 1999, Porter 1995]. Porter states that a value chain is composed of activities which can be classified as either primary or support activities [Porter 1995].

- **Primary activities** represent the activities involved in the physical creation of the product and its sale and transfer to the customer, as well as after sales assistance. Primary activities include inbound logistics, operations, outbound logistics, marketing and sales and service.
- **Support activities** support the primary activities and each other by providing purchased inputs, technology, human resources and various firm-wide functions. They include procurement, human resource management, technology development and firm infrastructure.

Porter further suggests that competitive advantage often derives as much from linkages between activities as it does from the individual activities themselves [Porter 1995]. Linkages basically represent the relationships between the performance of one value activity and that of another. Linkages can thus refer to relationships between the activities in the organisation itself and to the links that the organisation has with its suppliers and customers. Porter identifies each of these links as a source of potential competitive advantage, for they represent the relationships between the way one value activity is performed and the cost or performance of another. Porter states that linkages are found within the activities occurring in the organisation itself and within the links that the firm has with its suppliers and customers and that each of these links is a source of potential competitive advantage. By focusing on specific links, an organisation can gain a competitive advantage in the related business process. In today's terms this competitive advantage is achieved using Internet and eCommerce technologies and in particular, B2B and B2C approaches.

The challenge to manufacturing today is to facilitate inter-enterprise networking across the value chain and to gain competitive advantage by using information technology to achieve efficient linkage across the value chain.

3.4.3 Extended Enterprise and the Total Product Cycle View

In supporting sustainable manufacturing, it is important that the manufacturing system model show:

- The flow of goods and information up the point where the product or service is sold to the customer.
- The recovery of resources (i.e. products, components, material and/or energy).
- The end of life products.

The extended enterprise model of manufacturing systems caters to these requirements (Figure 3.9). In resource recovery, once a product reaches its end of life, the customer makes this product available to a collector, who delivers it to a resource recovery service provider. The resource recovery service provider assesses the product for recovery under the following guidelines:

- The product may be remanufactured.
- Some or all of the components may be reclaimed for use in other products.
- Some of the base materials may be reclaimed.

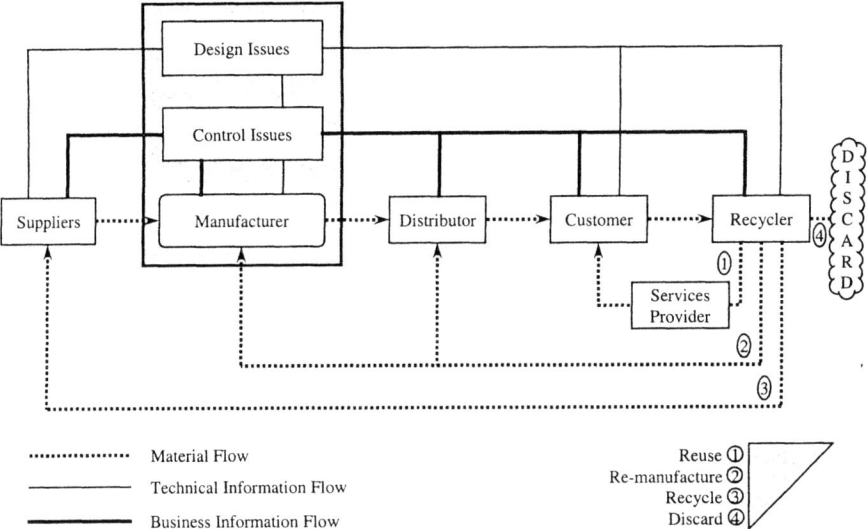

Figure 3.9 Extended enterprise including product recycling

If no recovery is possible, then the product, or part of the product is disposed of (e.g. some product components may be reclaimable while others may be disposed and some base material may be reclaimed). After product recovery assessment, the resource recovery service provider may use a distributor to sell products, components and base materials on to external customers or may return the components to business units within the supply chain. The repair and maintenance service provider uses new and used components and base materials to service and maintain in-use products.

The technical information associated with resource recovery flows from the resource recovery service provider to the design and marketing departments. This information provides design guidelines for the design for environment (DFE) i.e. how the product may be designed, so that at the end of its useful life, its constituent subassemblies, components and materials can be reclaimed and relevant systems are in place for their reuse.

3.4.4 Porter's Model and the Extended Enterprise

Porter suggested that the competitive forces, which operate on any organisation, include supplier power, buyer power, potential entrants' threat, substitute products threat and existing firms' rivalry. The nature of these forces becomes modified within the context of the extended enterprise:

- **Supplier power**. In Porter's model an adversarial relationship exists between suppliers and buyers. However, in the extended enterprise model it is vital that a harmonious relationship exist between the suppliers and the other partners of the enterprise, due to their mutual interest in bringing the product/service to the market on time and within budgetary constraints.
- **Buyer power**. Porter suggests that buyers exert power by forcing cost reductions and bargaining for higher quality products and/or services. In an extended enterprise setting, due to the buyers' close relationship to the customer, buyers will be more receptive to the needs of the customers for a specific product and/or service and will 'assemble' an enterprise to satisfy these needs.
- **Potential entrants threat**. The extended enterprise has enhanced competitive abilities in comparison to non-extended enterprises. Thus, the threat of new entrants will primarily come from other extended enterprises.
- **Substitute products threat**. Whilst the threat of substitute products will always exist, as a result of the attractiveness of the products offered by extended enterprises, this threat will be reduced.
- **Existing firms' rivalry**. Rivalry will still exist between enterprises in an extended enterprise environment.

3.5 The Virtual Enterprise Framework

Unlike the extended enterprise formation where, more often than not, existing supplier/buyer (or supply chain) relationships are strengthened through the use of ICT technologies, the availability of ICT technologies can also initiate the formation of entirely new enterprise networks.

The virtual enterprise is one manifestation of the organisational response to the globalisation of today's markets. The baseline for a virtual enterprise is the customer needs. These needs can be extensive and unique (e.g. a large project based contract) or small but with numerous variations. For example a number of complementary companies specialising in the repair and maintenance of household items may form a virtual enterprise to give a comprehensive service to its potential customers. These services might include the maintenance of a house structure, all forms of energy supplies, telecommunication and entertainment links, repair/maintenance of household durables like cookers, washing machines, refrigeration equipment and the recycling and waste disposal. Each of these services will require unique core competencies. Thus, several small specialist companies can increase their potential customer base by pooling their competencies. The attractiveness for the customer of such an enterprise will be that there will be only a single contact point for most of the household related problems.

From the above scenario, it is clear that the pooling of more than one core competency is necessary for the formation of a virtual enterprise. Taken literally, there is nothing new in small enterprises joining forces to strengthen their marketability. However, it is the availability of ICT technologies that has given small enterprises entirely new platforms to collaborate efficiently to supply an effective product/service. Thus, it should be noted that ICT is an enabler technology and a necessary (though not sufficient) condition for a virtual enterprise to exist.

Business partners in virtual enterprises can retain the individual agility of the consortium members to undertake their own business operations and quite possibly participate in other EE or VE type of projects simultaneously [Jagdev 1998, Hunter, 1999]. As information and communications technologies overcome the constraints of time and distance, it has become possible to create virtual organisations, where independent companies, temporarily linked by ICT networks, share skills, costs, and can access each other's markets.

Zhao *et al.* [1999] define a virtual enterprise as follows: "a virtual enterprise is a term often used to describe such collaborations when the participating organisations are distributed over large geographic areas and come from multiple independent companies. A virtual enterprise allows its member organisations to respond collectively to favourable market conditions for new product development in situations where the members individually would be unable to respond effectively". In principle, small and medium size companies participating in a virtual enterprise, get access to the resources of a large organisation while retaining the agility and independence of a small one. Skyrme [1996] suggests that the following benefits may be obtained through the construction of a virtual enterprise:
- Access to a wide range of specialised resources,
- Presentation of a unified face to large corporate buyers,
- Individual members retain their independence and continue to develop their core competencies,
- Reshaping of the enterprise and members according to the project or task in hand,
- Lack of need to worry about 'divorce settlements' as in formal joint ventures.

Davidow and Malone [1992] echo this sentiment when they suggest that networks of co-operating manufacturers offer the best strategy for structuring and revitalising the corporation for the 21st century. They name these networks 'Virtual Corporations'. In their book entitled 'Virtual Corporation' [Davidow 1992] they present their vision of 21st century industries, which will be built around "a new kind of product, delivering instant customer gratification in a cost-effective way". These products have a very rich service component, that is often more important than the tangible characteristics of the product. They must and can therefore be produced in several locations and offered in a great number of models and formats. Davidow *et al.* call these products 'virtual products'. They believe that a manufacturing company will not be an isolated facility of production, but rather a node in the complex network of suppliers, customers, engineering, and other service functions. The real time adaptation of the virtual product to the customer needs requires the virtual corporation to maintain integrated and ever changing data files on customers, products, and production and design methodologies. They therefore speak of it in terms of patterns of information and relationships that will appear less as a discrete enterprise and more as an ever-varying cluster of common activities of suppliers and their downstream customers in the midst of a vast fabric of relationships. These relationships will be built on principles such as 'joint destiny, trust and sharing information'.

The virtual corporation of Davidow and Malone is described as almost edgeless, with permeable and continuously changing interfaces between company, supplier and customers. Nevertheless, Davidow and Malone stress the importance of brand names and product identity. A virtual corporation is identified by the activities carried out and the products delivered. In fact, a virtual corporation is defined through the product or product line it produces.

Therefore, a *virtual enterprise can be defined as a network of independent organisations that jointly form an entity committed to provide a product or service.* This is due to the fact that, from the customer's perspective, these independent organisations, for all practical and operational purposes, are *virtually* acting as a single entity/enterprise. Taken from this perspective, the following can be identified as major characteristics of the virtual enterprise:

- The partners (or more) in the virtual enterprises are individuals and independent companies who come together and form a *temporary* consortium to exploit a particular market opportunity.
- Within the scope of collaboration, partners share a vision and work towards a shared goal.
- Partners in virtual enterprises make extensive use of ICT technologies for communication and sharing of information. Most of the day-to-day information exchange among the partners will almost always be automatic and without human interference.
- Virtual enterprises assemble themselves based on cost effectiveness and product uniqueness without regard to organisation size or geographic location.
- Unlike Supply Chains or Extended Enterprises, virtual enterprises once formed will have a unique dynamic, new identity and quite possibly a new name.
- The efficiency of the virtual enterprise is greatly determined by the speed and efficiency with which information can be exchanged and managed among business partners. Efficient collaborative engineering, production and logistics require effective electronic management of engineering and production information. Thus it is a prerequisite that participating enterprises have sufficiently sophisticated IT and decision support tools and mechanisms to make the integration possible.
- Virtual enterprises pool costs, skills, and core competencies to provide world-class solutions that could not be provided by any one of them individually. Therefore, virtual enterprises often focus on complete products or solutions as opposed to providing partial solutions in a value chain.
- The decisions are jointly arrived at by making the best use of the competencies among the partners.
- Virtual enterprises will often be complex networks where each enterprise can be seen as a node.

- The relationship between a set of nodes in a virtual enterprise will mostly be non-hierarchical in nature.

3.6   The Extended Products Framework

Just as individual enterprises work together in extended and virtual enterprises so also are many products migrating into extended products by: combining traditionally separate products and services, responding to demands for new services and embedding new services (frequently information and communications technology based) into traditional products.

Core products are 'layered' with additional services to make the overall package more attractive to the prospective customer. This aggregation, which we term *Extended Products*, consists of tangible core (manufactured) products and additional intangible frequently service-based components. Furthermore, the development of *Extended Products* is not only driven by competitive pressures to respond to market demands; legislative pressures may also serve to give rise to Extended Products. For example legislation in the European Union borne out of environmental concerns, is shifting the responsibility for 'end-of-life' products from society at large back to the manufacturers and distributors of the products. As a consequence, new Extended Products concepts have evolved that package end-of-life take-back and associated recycling with the core product. A simplified perspective on extended products model is visualised in Figure 3.10.

Figure 3.10 shows seven key considerations that have direct impact on the formation of extended products. Similarly, when enterprise networks are involved in the formation of extended products - as it will often be the case - one has to look at all aspects of the life-cycles of the enterprise networks.

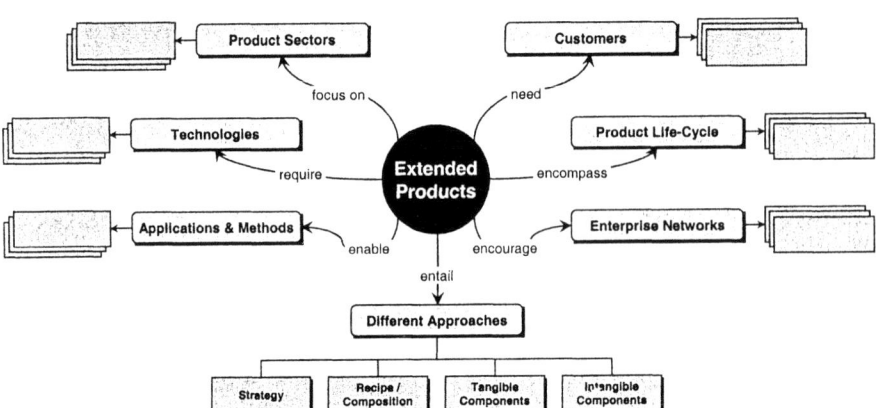

Figure 3.10 Extended enterprise including product recycling

Figure 3.11 gives a selection of issues partners have to consider for the smooth and successful operation of the enterprise network, in addition to the clearly defined (and understood by all parties) issues related to the formation and dissolution of the enterprise networks (see also Jagdev *et al*, 2001).

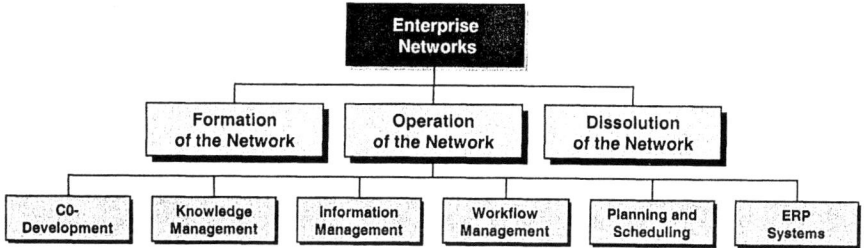

Figure 3.11 Further specification of the enterprise networks

One of the consequences of extended products is the development of marketing and sales concepts that delay assembly and the incorporation of associated services until final specifications are confirmed by individual customers. With the advent of the Internet, the customer catchments area has become global in nature and the scope for communications within the ancillary services suppliers (e.g. insurance/finance packages) has become seamless – thus facilitating the formation of enterprise networks constituting not only the suppliers of core products but also associated services. Indeed, the concept of the extended product is not restricted solely to manufactured goods. Consider the Marriott International Hotels web site [http://www.marriott.com], where all of the information is kept in a database and is presented to the site visitor according to the visitor's search string. This website describes perfectly an extension of the product (i.e. the hotel room) towards a comprehensive package of services for the benefit of the customer. The fundamental concepts behind the term extended products can be represented in the form of a layered model, as depicted in Figure 3.12.

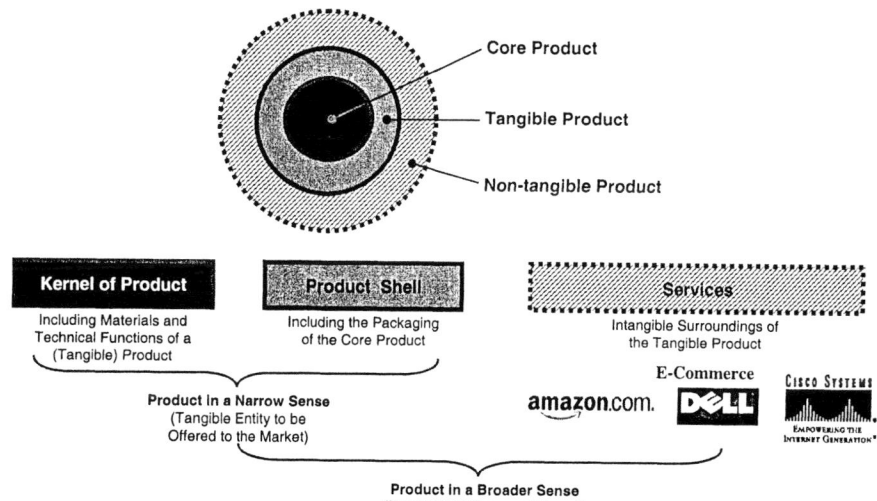

Figure 3.12 The extended product – product becomes a pure service

The three principal rings can be described as follows:

- The first (inner) ring describes the core product which is closely related to the core function (s) of a product. As an example consider the mobile phone which has as a core functionality, the ability to receive and transmit voice and data (i.e. communicate) in a location independent mode.

- The second (middle) ring describes the packaging of the core functions and includes tangible features of the chosen product. The features of the tangible product differ from manufacturer to manufacturer. In the case of the mobile phone, tangible features can include the packaging, the user-friendly nature of the software and the user interface.

- The third (outer) ring addresses the intangible assets of the product. Intangible assets surround the tangible product and in general are similar for similar products. Thus, various service providers provide a set of mobile phone services (news updates, information alerts from stock markets or sports fixtures, ability to book flights and hotel rooms etc.) In fact, such service providers are ready to subsidise the 'purchase' of the core product (the mobile phone) to facilitate the use of their networks and their information products.

Figure 3.12 also attempts to locate some well-known companies within these layers. A company such as AMAZON sells products over the internet offering key add-on services such as logistics, packaging, etc. DELL extends the added value to the customer by offering services within the complete product life cycle. The customer can purchase a PC, configured to meet his personal requirements, and have it delivered within a very short time.

The current trend, as depicted in Figure 3.13, is for an increasing share of value for the non-tangible compared to the core/tangible or in 'narrow sense' product element of the 'package'.

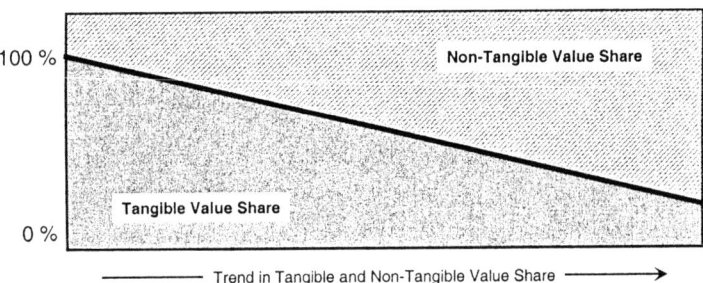

Figure 3.13 Increasing the non-tangible value-share

Just as the term extended enterprise comprises more than a single enterprise, the term extended product comprise more than the core or tangible product. The ring layered model of Figure 3.14 may be useful in articulating the notion of the extended product as a collection of services layered on to a core product and provided by a number of diverse service providers, who together ultimately seek to provide defined benefits to consumers.

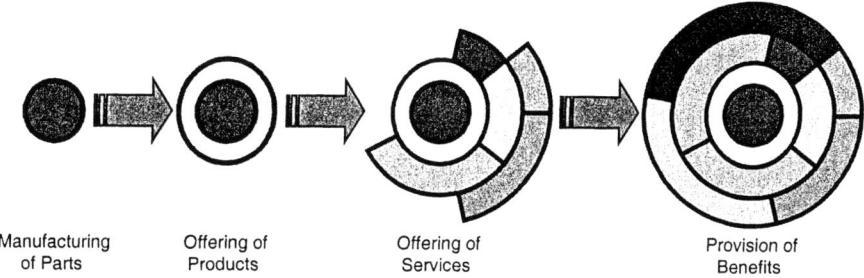

| Manufacturing of Parts | Offering of Products | Offering of Services | Provision of Benefits |

Figure 3.14 Changing the Focus: from manufacturing of parts to provision of benefits

3.6.1 Extended Enterprise Networks and Extended Products

As shown in Figures 3.12, 3.13 and 3.14 an extended product consists of tangible and intangible elements. This results in a new view of the product, emphasising the utility of the package (product and services) as distinct from the physical product itself. Extended or virtual enterprises are needed in order to provide the totality of services or intangible elements of the extended product over its life cycle. Extended or virtual enterprises frequently deal with the product over its total life, 'intervening' at various stages to provide a necessary or requested service. Figure 3.15 depicts this approach. For a particular customer an extended product will often include only a subset of the available services, based on individual customer requirements.

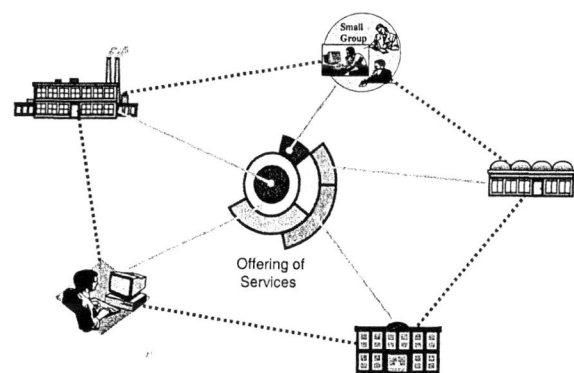

Figure 3.15 Extended products in the extended enterprise

Electronic hubs (Figure 3.16) offer a new platform for trading goods and services [Gates 2000]. Earlier we suggested that extended products will almost certainly be offered by extended or virtual enterprises. As providers of what are essentially market platforms, electronic hubs deliver an infrastructure which facilitates the ready availability of, and access to such products.

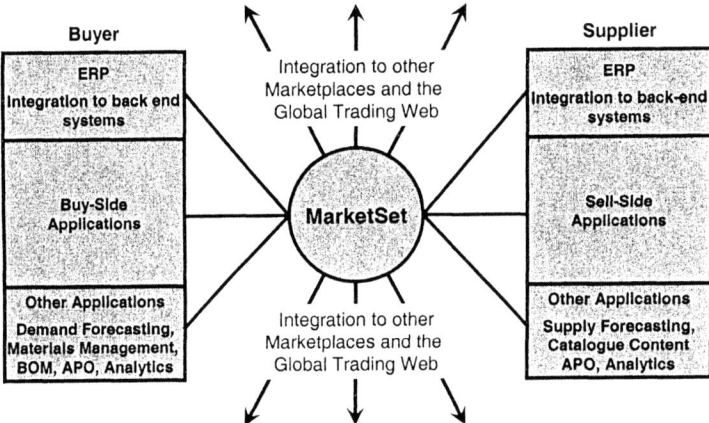

Figure 3.16 A typical hub perspective: CommerceOne's MarketSet platform hub [after Reily *et al.* 1998]

3.7 The Thinking behind the AMBIT Approach

In an environment where product differentiation and cost minimisation are critical, where extended enterprises involving co-operation across supply chains are becoming the norm and where the organisation must be aware of the business forces which impact it (i.e. political, social, economic and technological), information technology (IT) is of great importance. IT supports the rapid and accurate acquisition of a valuable organisational asset i.e. information. The most relevant information required by an organisation is that which supports strategic decision making. The decision to expand into new markets, to develop a new product, to invest in a new technology or a new programme of process improvement are all examples of strategic decisions. A key element of effective strategic decision making is access to reliable, accurate and timely information. However, even on the satisfaction of this requirement, a planning dilemma can arise due to the nature of the managerial work. Managers typically lack the time to study and analyse changes in consumer buying patterns and changing competitive positions, never mind having the time to learn the intricacies of information systems. Furthermore, the changeable nature of the organisational environment commonly prohibits the testing of theories and the evaluation of conflicting ideas. Whilst management can respond by adopting 'a hands on' learning attitude, they do not in general, work in an ideal environment where the impact of their strategic decisions may be readily assessed. Delays between strategic decision making, implementation and evaluation may be protracted, and as a result, strategic decision evaluation may become difficult. Unanticipated changes in consumer buying behaviour and/or economic conditions can also affect strategic decisions, especially if they are made over a long time span. In addition to the need to be vigilant with regard to current market trends and capable of evaluating the strategic implications of such trends, it is vital that managers are aware of advances in manufacturing technology and/or programme domains and are capable of assessing the strategic implications of new technology and/or new programme introduction on their organisation.

The ease of introduction and use of advanced manufacturing technologies and programmes is related to the organisation's ability to quantify the potential impact of such technology and programme introduction on its business. Given that investment decisions are necessary to support the organisation in its chosen market, it is vital that major investments in new manufacturing technologies and programmes be compatible with the organisational business goals. Brown agrees that "investment in technology is a strategic decision" [Brown 1996], which is crucial to organisational success (e.g. technology can affect key competitive factors such as cost, delivery speed and flexibility).

Hamaker [2003] states that organisations invest billions of dollars in "information technology (IT) yet researchers and decision-makers question the worth of those investments. Inquiries about the linkage between business performance and IT investments often puzzle researchers". Thus, since any strategic decision to invest in new manufacturing technologies and programmes can result in a large financial outlay and disruption to the organisation, it is vital that there be a technological and manufacturing presence at the senior management level within the organisation. The knowledge and experience of this presence can be used to guide the extent and applicability of strategic technology investment made at the corporate level. As Twiss *et al.* state "the contribution of technology is too important for corporate success to be left to the judgement of the technologist alone. The technological tail cannot be allowed to wag the corporate dog" [Twiss *et al.* 1989]. The decision to invest in new technologies should not be at the whim of a particular technical specialist, but should instead be a holistic decision for the organisation. Given the importance of strategic decision making, it is critical that decision makers be aware of the organisational implications of the decisions they make. If the strategic decision maker knows little or nothing about manufacturing, then it is quite likely that the strategic decisions made by him will be of a detrimental nature to the organisation.

In addressing this situation, Steve Brown proposes the following two alternatives [Brown 1996]:
1. The ultimate decision-maker is the executive who is familiar with manufacturing issues, in addition to other business considerations.
2. The ultimate decision-maker is not knowledgeable about manufacturing but has support, in his decision making task, from the relevant personnel (i.e. manufacturing managers) within the organisation.

However, a third proposal is the following: *the ultimate decision maker is the manager or team of managers, who use a strategic decision support tool as a means of supporting the strategic decision making process. The development of such a strategic decision support approach is the ultimate objective that AMBIT seeks to address.*

The dilemma to invest in a new manufacturing technology and/or programme is not an issue per se, as most organisations must invest in order to remain competitive. However, what is an issue is the type and extent of such investment. How does an organisation quantify success in new technology and/or programme investment, given that the effect of the investment choice is rarely either obvious or straightforward?

The traditional approach, where senior managers used their intuition is of little use. Far too often, organisations have invested heavily in a technology and/or programme which has resulted in little or no benefit. Clearly, what is required is a managerial tool, which supports the manager in his strategic decision making responsibility; a tool which facilitates rather than replaces thinking; a tool which serves the existent process rather than seeks to exert its own imperatives and control. In Figure 3.17 the formalisation edge for such a tool is depicted. This figure clearly illustrates the types of support formalisation which build up to an edge, and then drop off, once support becomes control.

"Formalisation can pertain to time, to location, to participation, to agenda and to information as well as, but with only the greatest of care, to process itself. It can help to focus attention, stimulate debate, keep track of issues, promote interaction and facilitate consensus. As Langley points out "ideas don't come out of planning ... ideas are in the air. But the plan will force us to make an effort to group things together and to define these orientations more clearly. I don't think the plan will be a surprise. For most people, it's just a chance to articulate their ideas. I've done research work and I think there's an analogy here. At some point, you have done a lot of work and collected a lot of data. Then the time comes when you have to present it somewhere ... don't do anything new, but it forces you to put the data together, to synthesis it ... to discuss things based on the synthesis. And often, just getting the data together generates new ideas". [Mintzberg 1994]

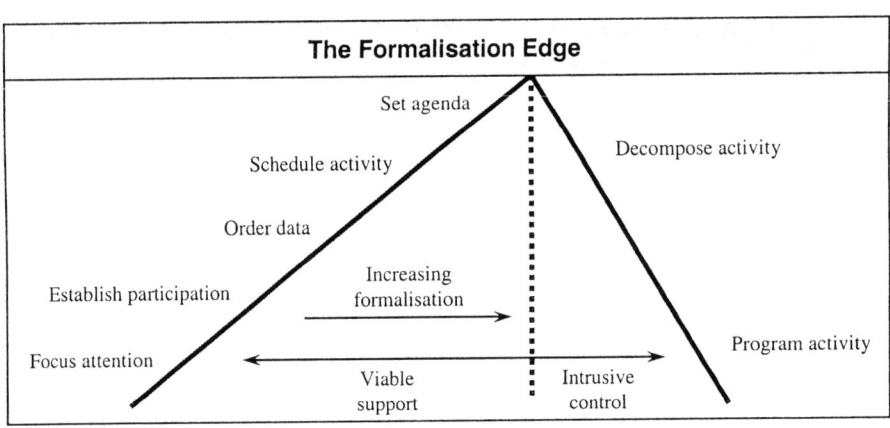

Figure 3.17 The formalisation edge [after Mintzberg 1994]

The AMBIT approach fulfils the support formalisations (i.e. focus attention, establish participation, order data, schedule activity and set agenda) required by managers. It does so by facilitating rather than replacing thinking and by allowing the user to explore different possibilities and options rather than identifying a definitive path. In developing the AMBIT approach the needs of managers and management teams, in addition to the need to minimise intrusive control, have been taken into consideration.

The AMBIT approach is developed to assist a manager or a team of managers in deciding whether or not to:
▪   Invest in changes to an existent technology and/or programme.
▪   Invest in a new technology and/or programme.

The AMBIT approach supports the manager or team of managers in assessing the strategic impact of such technology and/or programme changes and/or new implementations.

3.7.1   AMBIT Manufacturing Model

In order to provide the basis for organisational strategies and methods, it is necessary to define an explicit manufacturing model for the AMBIT approach, a model which satisfies the needs of today's manufacturing organisation. The AMBIT model for a manufacturing organisation (based on the extended enterprise model for a manufacturing system) is depicted in Figure 3.18.

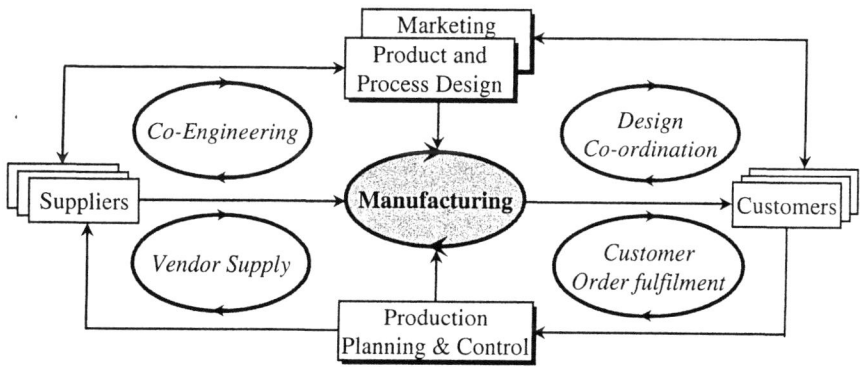

Figure 3.18 The AMBIT manufacturing model[9].

This model defines the main activities occurring within a manufacturing environment. It essentially represents a functional view of a manufacturing organisation. The horizontal axis of the AMBIT manufacturing model highlights the linkage between suppliers, manufacturing and the customers. As it characterises the flow of product and material from the suppliers through manufacturing to the customers, it emphasises the forging of closer links between suppliers, customers

---

[9] We have used this simplistic model of the Extended Enterprise for the AMBIT. We could equally have used a more sophisticated model, for example Figure 3.8 earlier that includes the recycling.

and the manufacturing system. The vertical axis illustrates the linkage between product and process design, manufacturing, and production planning and control. These operational elements, which represent the traditional CIM elements of the manufacturing environment, are instrumental in ensuring that the activities in the horizontal axis are co-ordinated and supported. The direction of the arrow flows in the diagram indicates the flows between these elements.

Based on the generic operational elements (i.e. supplier(s), product and process design, marketing, customers, manufacturing and production planning and control), the associated macro business processes include:
- Design co-ordination process.
- Customer order fulfilment process.
- Vendor supply chain process.
- Manufacturing process.
- Co-engineering process.

All of the operational elements and their interactions through the macro business processes impact the overall objectives of the organisation in terms of its financial and market position. This position is in turn set by, and should satisfy, the organisation's corporate strategy.

Linkages within the AMBIT Manufacturing Model

As already outlined, Porter suggests that competitive advantage frequently derives as much from linkages between activities, as it does from the individual activities themselves [Porter 1995]. Linkages refer to relationships between the activities in the organisation itself and to the links that the organisation has with its suppliers and customers. Porter's concept of linkages is depicted in the AMBIT manufacturing model (Figure 3.18) by the arrows. Each of the macro business processes contains several different linkages. As an example, linkages associated with the vendor supply chain process include ordering materials from suppliers and receiving these materials into the manufacturing process. By focusing on these linkages, an organisation can gain a competitive advantage in the supply chain loop. Similar linkages, offering similar opportunities for gaining a competitive advantage, exist within each of the other processes.

The AMBIT Manufacturing Model and the Value Chain

> The value chain brings out the point that business activities require the use of resources - human, material and capital assets - very often, as well as less tangible items such as knowledge, information and routine.
>   – Graham *et al.* 2000

Porter suggests that a value chain is composed of value activities, which are classified as either primary or support activities. Primary activities refer to the physical creation of the product and its sale and transfer to the customer as well as after sale assistance. Support activities support the primary activities and each other by providing purchased inputs, technology, human resources and various organisational wide functions [Porter 1995]. Porter's primary activities are reflected in the different elements within the AMBIT manufacturing model and the corresponding macro business processes. However, one predominant difference exists between Porter's value chain model and the AMBIT manufacturing model. Porter's model does not adequately account for the design co-ordination or co-engineering processes, apart from their consideration in the technology support activity. This represents a weakness in Porter's model, as the design co-ordination and co-engineering processes are vital to many manufacturing organisations. As an example, in an engineer to order environment, design co-ordination and co-engineering, are sources of major competitive advantage.

3.8   Overview of the AMBIT Approach

> Investment in capital equipment, leading edge technology and the development
> of skills creates a virtuous circle which drives up performance.   – DTI 2002

Corporate strategy and business strategy impacts on all operating aspects of an organisation; including marketing, manufacturing, finance, R&D, etc. The focus of the AMBIT approach is limited to the manufacturing operations of an enterprise. The majority of organisations traditionally dealt with manufacturing decisions as a series of technical problems that could be resolved one by one without undue consideration of their influence on the overall business strategy. The AMBIT approach addresses this situation by supporting the linkage of the business model to a manufacturing strategy model, so that the impact of a new technology and/or programme investment on the manufacturing strategy and ultimately on the business and corporate strategy can be evaluated. Simply put, the AMBIT approach responds to:

1. *The need to develop a statement of manufacturing strategy consistent with the business strategy.*
    Far too frequently, organisations have not had sound manufacturing strategies that are consistent with their overall business strategies [Mellor 2002, Platts *et al.* 1992, Mohd 2002]. Skinner claims that due to manufacturing often being ignored, the corporate strategic options become limited and concrete facilities, people and equipment become fixed into a non-competitive pose. Such a pose can impede innovation and progress. Whilst the importance of the manufacturing strategy to organisational success is clear, it is vital that the manufacturing strategy be integrated with the overall corporate strategy. In response to the need for such integration, the AMBIT approach has been developed to support the linkage of the business model to a manufacturing strategy model. In so doing, the high-level business objectives of a manufacturing organisation, which are generally abstract and generic in nature, are decomposed into operational level measures of performance. This decomposition is achieved through several steps, such as the generation of strategic performance indicators (SPIs) and measures of manufacturing performance (MOMPs). Chapter 4 discusses this decomposition process and the generation of MOMPs in detail.

2. *The need to assess the impact of new technology and programme introduction on the organisation from the viewpoint of employees, skills and manufacturing systems design and operation.*
    Management decisions to invest in manufacturing technologies and programmes are vital in equipping the organisation to be more competitive. Poor technology investment decisions can result in wasted funds and can damage an organisation's capability to satisfy customer demands. It is thus important that organisations be capable of evaluating the impact of manufacturing technology and/or programme investment from the viewpoint of employees, skills and manufacturing systems designs. The AMBIT approach has been developed to support the assessment of new manufacturing technology and/or programme introduction on the identifiable MOMPs.

3. *The need for a tool which fosters managerial learning and facilitates managerial interaction.*
    The AMBIT approach can provide a set of tools to capture and represent an individual company's understanding of the impact of new technology and/or programme introduction and/or change on their business. In order to optimise managerial interaction, it is important that:
    -   The learning profiles of the user(s) be taken into consideration. In Chapter 7, the relationship between creativity and learning, and decision making and learning is discussed.
    -   The AMBIT approach based tool should operate within a computer based learning environment (CBLE). A CBLE provides decision makers with greater scope of learning through conceptualisation, experimentation and reflection. Chapter 7 describes learning profiles and CBLEs.

In addressing these needs, the AMBIT approach is broken into three stages which are linked by a common set of measures termed MOMPs. MOMPs represent measures of manufacturing performance. MOMPs are extremely important because in addition to measuring performance, they also tend to dictate behaviour i.e. what you manage you can measure and what you measure is what you get [Chan *et al.* 2003, Boone 1991]. Given that the success or failure of a manufacturing decision is generally reliant on the right performance measure to support and drive it, the basis of most decision-making systems is driven by performance measurements (Figure 3.19).

The AMBIT approach allows management to gain an appreciation of the strategic importance of new manufacturing technology and/or programme introduction. It does so by supporting the strategic assessment of changes in manufacturing technologies and/or programmes on the strategic business metrics (via the measures of manufacturing performance).

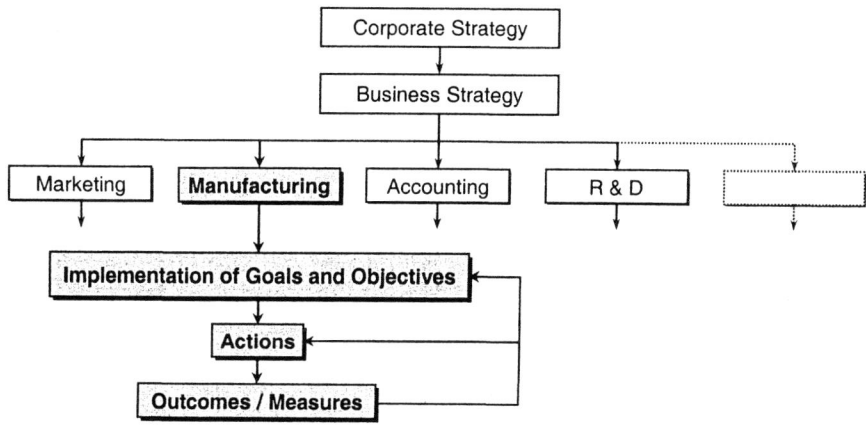

Figure 3.19 Position of performance measurement in implementing manufacturing strategy

The performance measurement framework, which forms a fundamental part of the AMBIT approach, is representative of a contemporary performance measurement system. It will be shown in the following chapters that the AMBIT approach satisfies the necessary requirements for an effective performance measurement system (see Chapter 2), i.e. the need for [Chan *et al.* 2003, Kennerley 2002, Waggoner *et al.* 1999, Barth 1996]:

- A direct relationship to the mission and objectives of the organisation.
- Validity.
- Accuracy.
- Enhancing motivation and communication.
- Completeness.
- Diagnosing problems.
- Reliability.
- Understandability and usability.
- Flexibility.
- Adaptability.
- Supporting the understanding of the organisation's progress and the current competitive position of the organisation.
- Maintainability.

3.8.1 An Outline of the AMBIT Approach

In the AMBIT approach, the focus is primarily at the functional level of the organisation. Thus, before using this approach, it is assumed that the business plan has been prepared and the CSFs (critical success factors) associated with this plan have been identified (Figure 3.20).

CSFs represent the key areas in which an organisation's performance must be satisfactory, if that organisation is to achieve and maintain competitive advantage.

These CSFs are then used by the AMBIT approach to provide a range of solutions which are compatible with the business plan.

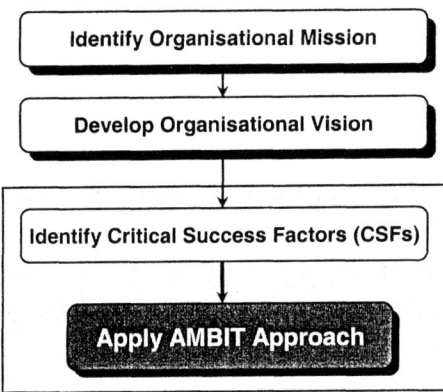

Figure 3.20 Scope of AMBIT approach

Critical success factors will differ both among industrial sectors and for individual firms within a particular sector. CSFs are generally sourced from industry-based factors, competitive strategy, industry position and geographic location, environmental factors and temporal factors. Once the CSFs have been ascertained, the AMBIT approach can be applied to achieve the desired changes in the manufacturing operations. The AMBIT approach comprises the three stages as depicted in Figure 3.21. We will now briefly discuss each of these stages. Chapters 4, 5 and 6 following will give a detailed treatment of each of these three stages.

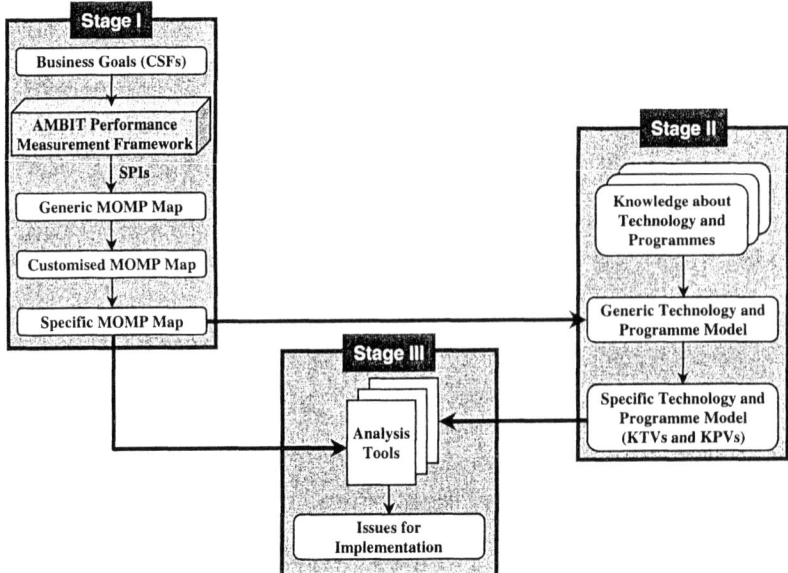

Figure 3.21 The AMBIT approach

Stage I (Chapter 4) of the AMBIT approach is concerned with the translation of the business plan/strategy into a specific MOMP map. As already mentioned, the business plan is expressed in the form of CSFs. However, because of the high level nature of the CSFs (i.e. they are composed of hard elements e.g. output, and soft elements e.g. working culture) they are often quite difficult to quantify. This also means that the task of relating them to the manufacturing strategy is also difficult. But why is it necessary to relate the CSFs to the manufacturing strategy? If the organisation is involved in some aspect of manufacturing, it only stands to reason that compatibility should exist between the business strategy and the manufacturing strategy, for that organisation to survive and prosper. Thus, to support such compatibility, what happens next is the decomposition of the CSFs into more concrete milestones against which the organisation's progress (based on its business strategy) can be tracked. These more concrete milestones are called SPIs (strategic performance indicators). However, due to the high level nature of the SPIs, the problem of quantification (explained in greater detail in chapter 4) still exists. As a result, the SPIs are decomposed into a generic MOMP map. A MOMP map is a set of measures of manufacturing performance (MOMPs). Customisation of this generic MOMP map results in the specific MOMP map, which ensures that the MOMPs are unique to the organisation in question. These specific MOMPs represent the key measures that are consistent with the overall business strategy.

In synopsis, Stage I is concerned with the translation of the business plan/strategy (expressed in the form of the business goals) into CSFs. These CSFs are in turn, expressed in the form of SPIs which are further expressed in the form of generic MOMPs (i.e. the generic MOMP map). The customisation and instantiation of these generic MOMPs, results in the generation of the specific MOMP map. This map reflects the unique characteristics of the organisation in question and clearly defines the manufacturing related measures that must be addressed in order to achieve the goals set by the business strategy. Therefore, we can say that the specific MOMP map links the business strategy to the manufacturing strategy to be adopted by an organisation.

Stage II (Chapter 5) of the AMBIT approach is concerned with the technology and the programme models. As already outlined, the goal of the AMBIT approach is to provide a mechanism by which changes to an existent (or the implementation of a new) technology and programme, affect the manufacturing and ultimately the business strategy. Thus, it is necessary to link the specific MOMPs (derived in Stage I) into low level technological and programme elements, which are easily quantified. This linkage means that changes to these technological and programme elements can be rippled backwards to see how such changes affect the specific MOMPs (i.e. the manufacturing strategy) and then in turn the SPIs and ultimately the CSFs (i.e. the business strategy). In the case of a technology, the quantifiable elements are termed key technology variables (KTVs), whilst in the case of a programme, they are referred to as key programme variables (KPVs). The mapping of the specific MOMPs onto the KTVs and the KPVs represents the generic technology/programme model (Figure 3.21). Meanwhile, the customisation of the KTVs and the KPVs results in the generation of the specific technology and programme model. This model reflects the unique technological/programme characteristics of the organisation in question.

Stages I and II are linked to each other through the set of specific MOMPs, which act as a bridge between the two stages. Once Stages I and II are complete, a set of analysis tools are used to relate the "specific MOMP map" to the "specific technology and programme model".

Stage III (chapter 6) of the AMBIT approach deals with the analysis of the new technology and programme investment on the manufacturing and ultimately the business strategy. The output from Stage III is a set of implementation issues.

Figure 3.22 summarises the interaction among these three AMBIT stages.

The advantage of the AMBIT approach is the fact that the ultimate decision maker, who might not necessarily know a great deal about manufacturing, will still be able to assess the effect of a technology and/or programme investment decision on the organisation. He will be able to evaluate the impact of the investment decision from the viewpoint of human resources and skills, organisational structures and resource requirements.

Use of the AMBIT approach supports the prediction of the future, strategic thinking and strategic decision-making, management team learning and enhanced managerial and management team creativity.

Some of the goals of the AMBIT approach are as follows:
- To promote organisational efficiency by allowing the user(s) to strategically assess the result of changes to an existent technology and/or programme and/or the introduction of a new technology and/or programme.
- To support the articulation of a business and manufacturing strategy which defines:
  o The product set of the organisation.
  o The market(s) at which the organisation is directed.
  o The technology, knowledge and skills required to satisfy the requirements of that market.
- To facilitate the planning of new manufacturing technologies and programmes (i.e. business process reengineering, concurrent engineering, lean production, etc.) introduction and to support an understanding of their impact in terms of human resources, skills, structures, manufacturing capacity and floor space requirements etc.

The AMBIT approach (and a possible AMBIT tool set, which could be built using the AMBIT approach template) will support management in focussing on its own core competencies and improving them if necessary.

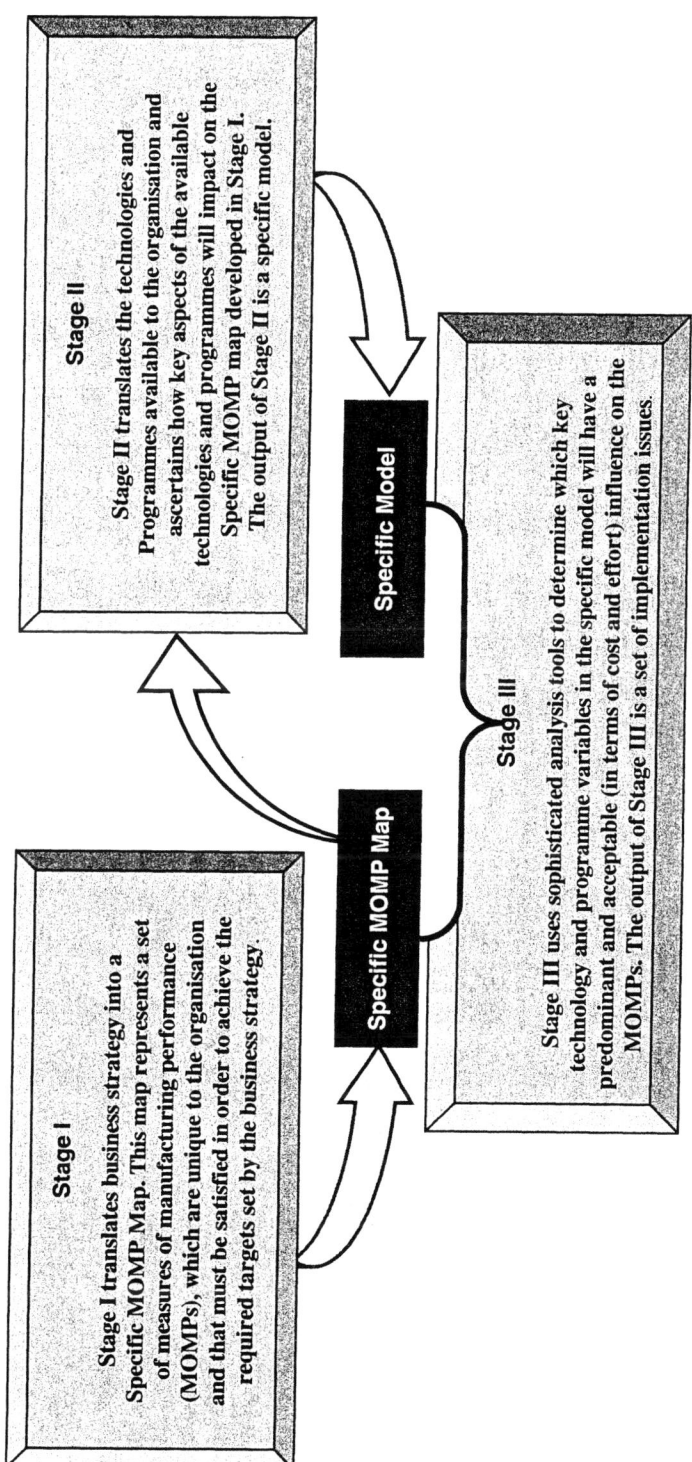

Figure 3.22 Interaction between three stages of the AMBIT approach

3.9   Conclusion

Today's manufacturers see the need to integrate the manufacturing strategy with the business strategy and ultimately the corporate strategy. With the growing focus on technology as a competitive weapon, appropriate technology investment decisions are vital to organisational success. Technology decisions can either support the organisation in its pursuit of manufacturing excellence, or result in large organisational liabilities.

What is required is a framework which accelerates the take-up and use of new manufacturing technologies and programmes by:

- Supporting management in understanding the medium and long-term organisational implications of technology and programme introduction.
- Providing management with both an approach and the necessary tools to:
  o   Analyse the competitive position of the organisation.
  o   Support the assessment of the technology and programme impact on selected business metrics.
  o   Plan the implementation of technologies and programmes.   .

The AMBIT approach, in its linkage of the business model to a manufacturing strategy model caters to these needs. The AMBIT approach is a strategic decision support tool developed to assist a manager or a team of managers in deciding whether or not to:

- Invest in changes to an existent technology and/or programme.
- Invest in a new technology and/or programme.

In this chapter, the importance of the manufacturing function and its relationship to the business strategy was emphasised. Furthermore the manufacturing function was considered in the context of modern developments in manufacturing systems, including the emergence of extended and virtual enterprises and extended products. Such developments, in so far as they suggest a more complex environment for manufacturing enterprises and therefore manufacturing decision makers, inform the development of the AMBIT approach. Indeed the business processes used in the AMBIT performance measurement framework to be discussed in the following chapter are based on an extended enterprise model of manufacturing systems.

Chapter 4

The AMBIT Approach – Stage I

Keywords: Organisational mission, organisational vision, CSFs, AMBIT performance measurement framework, SPIs, MOMPs, customer order fulfilment, vendor supply, design co-ordination, co-engineering, technical systems, transformations, time, cost, quality, flexibility, environment, capital goods, commodities, fashion goods, durables, jobbing shop, batch production, repetitive production, mass production, customer order decoupling, engineer-to-order (ETO), make-to-order (MTO), assemble-to-order (ATO), make-to-stock (MTS), generic MOMP map, particularisation and specific MOMP map.

Chapter Objectives: The purpose of this chapter is to show the reader the linkage of the business goals, expressed in terms of the organisation's critical success factors (CSFs) to the manufacturing strategy by means of the measures of manufacturing performance (MOMPs). Having read this chapter, the reader should become familiar with:

- The AMBIT performance measurement framework, based on three axes as follows:
    - o The manufacturing business processes axis based on the extended organisation model of manufacturing systems.
    - o The macro measures of competitive performance axis based on five macro measures namely cost quality, flexibility, time and environment.
    - o A classification of manufacturing systems axis based on a typology of manufacturing systems. In our case we will use the customer order decoupling point based typology.
- The derivation of strategic performance indicators (SPIs) and associated measures of manufacturing performance (MOMPs).
- The development of the generic MOMP map.
- The customisation of the generic MOMP map.
- The derivation of the specific MOMP map.

4.1   Introduction

Stage I of the AMBIT approach is concerned with the translation of the business goals, represented as critical success factors (CSFs) into a specific measures of manufacturing performance (MOMP) map relevant to the manufacturing organisation in question. In Stage I, the strategy of a manufacturing organisation is expressed in terms of these MOMPs. This is achieved through the following five steps (Figure 4.1):

Step 1. Identify the critical success factors (CSFs).

Step 2. Develop the AMBIT performance measurement framework (also termed as the AMBIT Cube). This will result in the identification of strategic performance indicators (SPIs).

Step 3. Develop the generic measures of manufacturing performance (MOMP) map.

Step 4. Customise the MOMP map.

Step 5. Develop the specific MOMP map.

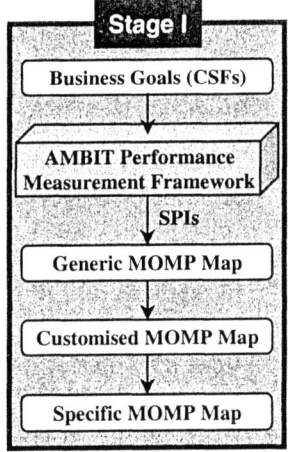

Figure 4.1 The AMBIT approach - Stage I

The key features of this stage of the AMBIT approach are the Performance Measurement framework and the MOMP map. They allow for the intangible and conceptual issues to be expressed in more concrete and tangible forms. For example, a business strategy of 'Overall Cost Leadership' is understandably difficult to quantify, with regard to its effect on human resources and skills, organisational structures and resource requirements. However, when this business strategy is expressed in terms of the critical success factors, the AMBIT performance measurement framework allows for its translation into strategic performance indicators (SPIs). These SPIs in turn represent the more tangible and concrete form of the CSFs.

However, the problem of quantification still exists. Whilst the SPIs are more concrete than the CSFs, they are themselves difficult to quantify. For example, if the

SPI is 'Time Consumed by the Manufacturing Process', then how can a value for this SPI be determined for an assembly technology, without understanding the details concerning the physical implementation of that actual technology in the first place? What needs to be done, therefore, is to decompose this SPI into a set of supporting measures of manufacturing performance (MOMPs). In this example, some of the related MOMPs would be 'Queue Time of Sub-Assembly Process', 'Processing Time of Sub-assembly Process' and 'Transport Time of Sub-Assembly Process'. This decomposition continues until an element is reached which can be used as part of a set of simple elements to differentiate between technologies and/or programmes. This breakdown will be achieved in Stage II. In this chapter, the steps of Stage I (as presented in Figure 4.1) are described and discussed in some detail.

An interesting feature of the AMBIT framework is its correspondence to the modelling dimension of the CIM-OSA model. CIM-OSA (Computer Integrated Manufacturing - Open Systems Architecture) defines an integrated methodology which supports all phases of a computer integrated manufacturing (CIM) system life cycle from requirements specification, through system design, implementation, operation to maintenance. In the CIM-OSA architectural framework, three different levels of generality, encompassing *generic* models, *partial* models and *particular* models are recognised. Partial models are models which are applicable to a specific category of manufacturing organisation and hence the interest in typology or the categorisation of manufacturing systems. Partial models refer to models of a particular category or class of manufacturing system. Particular models on the other hand are specifically concerned with a particular or unique manufacturing organisation within that category. A sequential instantiation process supports the transition from the generic state to the partial state and finally to the particular state [AMICE 1993].

The mapping of the AMBIT framework to the CIM-OSA model is shown in Figure 4.2.

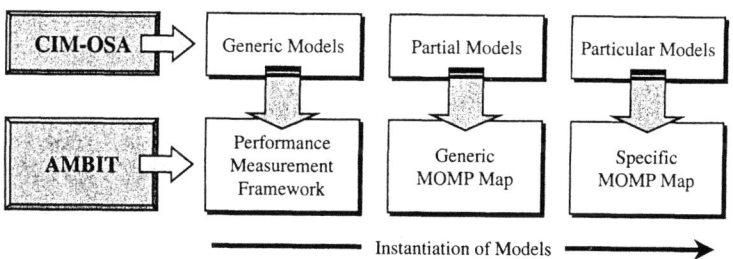

Figure 4.2 Mapping AMBIT framework to the CIM-OSA model

As indicated in this figure, the generic model is the AMBIT performance measurement framework which will be discussed in section 4.3. A partial model might involve the creation of a MOMP (Measurement of Manufacturing Performance) Map for an individual business process, in a certain category of manufacturing system e.g. assemble-to-order. (The details of various manufacturing classification systems or typologies will be discussed in section 4.3.3). A particular model would involve the detailed customisation of this MOMP map to reflect the specific situation in a particular or individual manufacturing organisation.

Using the structure suggested by Figure 4.1, we will now discuss the various steps in Stage 1 of AMBIT. The identification of business goals, expressed in terms of critical success factors is the first step.

4.2   Identify the Critical Success Factors (CSFs)

An organisation normally aims for coherence between its strategy (i.e. business direction) and its culture (i.e. the manner in which things are done within the organisation, the organisational policies and procedures etc.). Good alignment can support an organisation in moving forward in a clear fashion, whilst poor alignment can often result in the dissipation of organisational resources. In other words, poor alignment can result in the confused or non-aligned organisation (Figure 4.3).

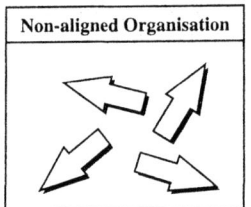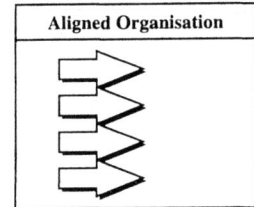

Figure 4.3 Non-aligned and aligned organisations [Mooney 1996]

In supporting organisational coherence, it is important that management be explicit in the definition of their:
-    Organisational mission (goals, purpose, beliefs).
-    Strategic aims (organisational priorities in the immediate to medium term).
-    Operational objectives (costed and timed action plans) and
-    Critical success factors (the factors which the organisation believes it has to get right if it is to satisfy its strategic plan).

For any organisation, the development of a clear vision and goals (and the detailed action plans which support these goals) provides strategic focus which in turn guides business decisions and individual actions. This single focus also helps to overcome rivalries amongst internal stakeholder and to release internal energy. It provides an environment where people become 'committed' to an agreed and well-articulated goal. It also provides a graphic picture of a possible future state – a state to which the organisation can begin to move [Mooney 1996]. However, given the far reaching benefits of a shared organisational vision, management in some organisations do not embrace the concept to the extent one would imagine. Hodgkinson interviewed 60 middle managers (from a large international corporation), concerning their views/insights on such a shared organisational vision. Her research showed that whilst these middle managers implicitly understood that a shared vision was vital in order to foster emerging organisational strategies which would have implications for future organisational success, their senior managers were not embracing the concept of a shared organisational vision [Hodgkinson 2002].

As already outlined, CSFs refer to the key areas in which the organisation's performance must be satisfactory, if the business is to thrive. The AMBIT approach operates under the assumption that the user(s) is familiar with the organisation's CSFs, so that the CSFs selected by him reflect the organisational focus. Examples of CSFs might be to shorten order delivery time and/or to increase delivery accuracy. Table 4.1 gives a small sample list of CSFs. Appendix to this book gives a selection of performance measures, many of which could be translated to CSFs.

Table 4.1 A sample list of critical success factors

A Selection of Critical Success Factors (CSFs)
▪ Reduce production costs
▪ Reduce the cost of the materials required to produce the product
▪ Improve the accuracy of the pricing information
▪ Improve the accuracy of the listing of all products
▪ Price products competitively
▪ Increase product quality
▪ Improve product reliability
▪ Increase the range of products
▪ Increase the range of product options
▪ Increase the range of environmentally friendly products
▪ Increase the number of recyclable products
▪ Increase the amount of environmentally friendly materials used in the production
▪ Reduce the time taken to introduce new products into the marketplace
▪ Reduce the time taken to design new products
▪ Reduce the cost of designing new products
▪ Increase the quality of the product design
▪ Increase the number of designs that can be recycled to form new designs
▪ Speed up the transmission of the design to and from the co-engineering team
▪ Improve the quality of the materials used in the production
▪ Reduce materials delivery times
▪ Improve material ordering procedures
▪ Improve the range of types of orders that can be placed with the vendors
▪ Shorten the time taken to deliver the materials required from the vendor
▪ Improve the accuracy of the listing of all materials
▪ Increase the accuracy of the order deliveries
▪ Improve materials ordering procedures
▪ Quick confirmation of having received the customer's order
▪ Increase greater ordering flexibility
▪ Reduce delivery times to supply the products to the customers
▪ Increase the completeness of the order when it arrives at the customer
▪ Meet delivery dates specified by the customer

Selected CSFs reflect the strategy employed by the organisation. CSFs, when further decomposed into Strategic Performance Indicators (SPIs), provide concrete milestones against which the progress of the organisation can be evaluated with respect to the satisfaction of its mission objectives and business goals. To derive

SPIs from the CSFs, the CSFs have first to be mapped onto the AMBIT performance measurement framework. The following section tackles this issue in some detail.

4.3   Develop the AMBIT Performance Measurement Framework

The AMBIT performance measurement framework provides a means of translating the business goals of a manufacturing organisation (expressed in terms of its critical success factors (CSFs)), into a specific MOMP (measures of manufacturing performance) map. Represented pictorially in Figure 4.4, the AMBIT framework shows the mapping of the CSFs in terms of the following dimensions (or axes):

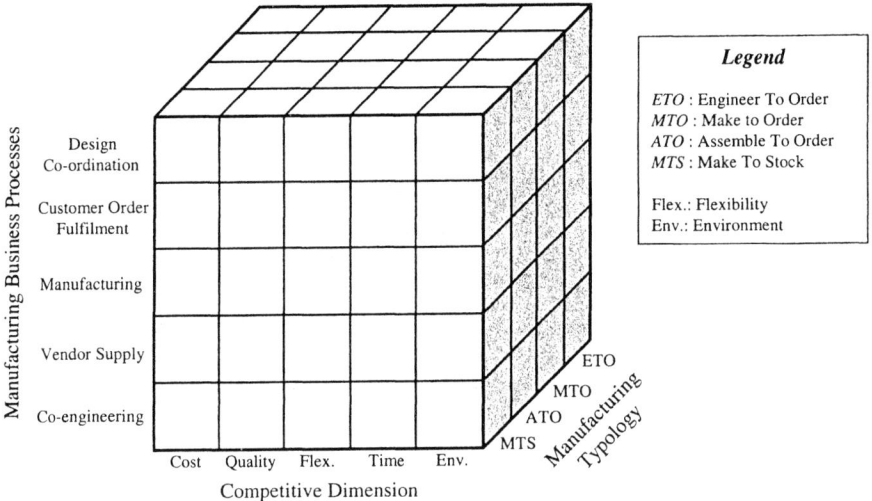

Figure 4.4 The AMBIT performance measurement framework

1. Manufacturing business processes. These processes reflect all of the activities occurring within a manufacturing organisation. Section 4.3.1 outlines how these processes are associated within the AMBIT manufacturing model.
2. Macro measures of competitive performance. These measures reflect the dimensions along which the organisation must compete in the marketplace and are discussed in detail in section 4.3.2.
3. Manufacturing typology. This allows the identification of the type of manufacturing organisation under study and thus forms the basis for the creation of the partial or sector specific models discussed above. The definition of the various sectors or partial models depends on the typology used to define the sectors. As Figure 4.4 suggests we will categorise manufacturing systems using the customer order decoupling point (CODP) typology. Clearly, other approaches could also have been used and these are discussed alongside CODP in section 4.3.3.

The strategy of a manufacturing organisation is expressed in terms of its critical success factors (CSFs). These CSFs are in turn used to identify corresponding strategic performance indicators (SPIs). Each point within the AMBIT cube, as defined by its co-ordinates, is termed a strategic performance indicator (SPI). SPIs can provide answers to such questions as: how is the organisation doing? or how will the organisational performance be evaluated?

An SPI is a business metric, which is a recognised measure of strategic performance. Given that the SPIs are derived based on a set of manufacturing business processes, macro measures of competitive performance, and manufacturing typologies, it is therefore important to:

- Define a generic business model which represents, at a high level, all of the activities occurring within a manufacturing organisation. This model should represent a functional view of the manufacturing organisation (see section 4.3.1).
- Select a set of performance measures, which represent the dimensions along which an organisation can choose to compete. These measures should reflect the ongoing changes in the manufacturing climate (see section 4.3.2).
- Determine a typology which defines both the sector in which the organisation competes and the unique operational characteristics of that sector (see section 4.3.3).

As an example, let us assume that the given business goal is to achieve 'Best customer value' (Figure 4.5).

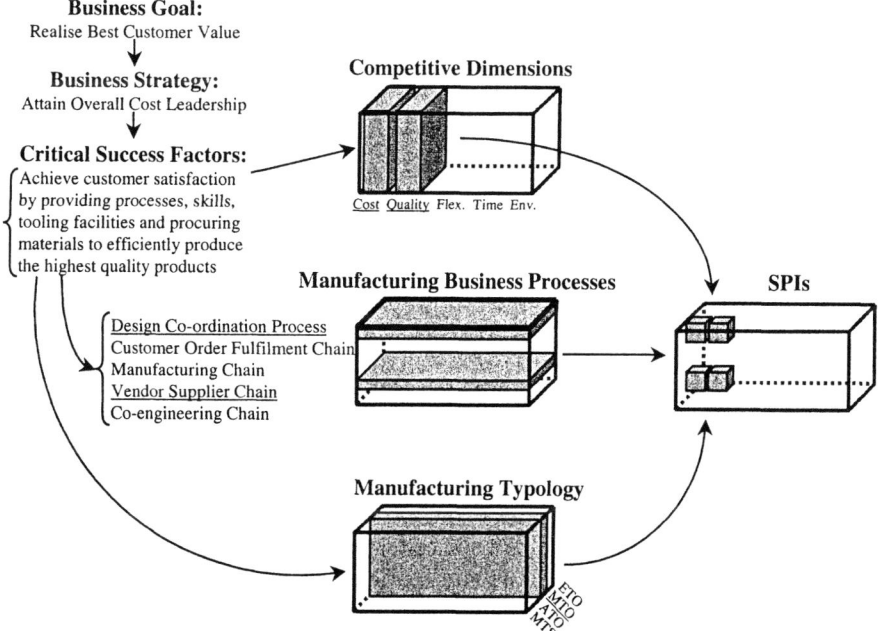

Figure 4.5 The elements of an SPI – a simple example

This goal might be expressed in terms of the business strategy as attaining 'Overall cost leadership'. The key area in which the organisation's performance must be satisfactory if the goal of overall cost leadership is to be realised is the following: 'achieve customer satisfaction by providing processes, skills, tooling facilities and procuring of materials to efficiently produce the highest quality products'. This represents the corresponding critical success factor (CSF).

However, it is difficult to relate this high level CSF to the strategy pursued by manufacturing. It is much easier to express it in more quantifiable terms i.e. in the form of strategic performance indicators (SPIs). Before it is expressed in the form of SPIs, the manufacturing business process(es), macro measure(s) of competitive performance and manufacturing typology corresponding to the CSF, need to be selected. This selection in turn results in the identification of the related SPIs.

In this example (as already mentioned), the key organisational area in which performance must be satisfactory, if the business is to be successful is the following: 'achieve customer satisfaction by providing processes, skills, tooling facilities and procuring materials to efficiently produce the highest quality products'. In other words, the organisation wants to be able to deliver high-quality, low-cost products to the customer. Hence, two of the manufacturing business processes that are affected by this CSF are the design co-ordination process and the vendor supply chain process.

The corresponding macro measures of performance are quality and cost. Finally, as the organisation in question maintains a stock of raw materials, so that following receipt of an order for a particular design, the product is manufactured from these components, the manufacturing sector involved is make-to-order (MTO). Consequently, the result of the decomposition of the CSF into related SPIs is as follows:

- Quality in the vendor supply chain (VSC) within a make-to-order (MTO) environment - VSC/Quality/MTO.
- Cost in the vendor supply chain (VSC) within a make-to-order (MTO) environment - VSC/Cost/MTO.
- Quality in the design co-ordination process (DCP) within a make-to-order (MTO) environment - DCP/Quality/MTO.
- Cost in the design co-ordination process (DCP) within a make-to-order (MTO) environment - DCP/Cost/MTO.

In synopsis, each SPI has three attributes: manufacturing business process, macro measure of competitive performance and manufacturing typology. Given that there are 5 manufacturing business processes, 5 macro measures of competitive performance and 4 manufacturing typologies from which to select, the number of SPIs which can be selected at any one time is 100 (5*5*4) SPIs. This set of strategic performance indicators represents a set of basic metrics, the elements of which are used to quantify the efficiency and/or effectiveness of an action.

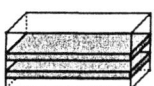

4.3.1　Manufacturing Business Processes Axis

The first axis of the AMBIT performance measurement framework is defined by a set of macro business processes (see AMBIT manufacturing model Section 3.7.1). As outlined in Figure 3.18, the AMBIT manufacturing model and the associated macro business processes evolved from the extended enterprise model of manufacturing systems. These macro business processes indicate the sequence of activities and flows between the five main operational elements. They also represent the first axis of the performance measurement framework (Figure 4.6).

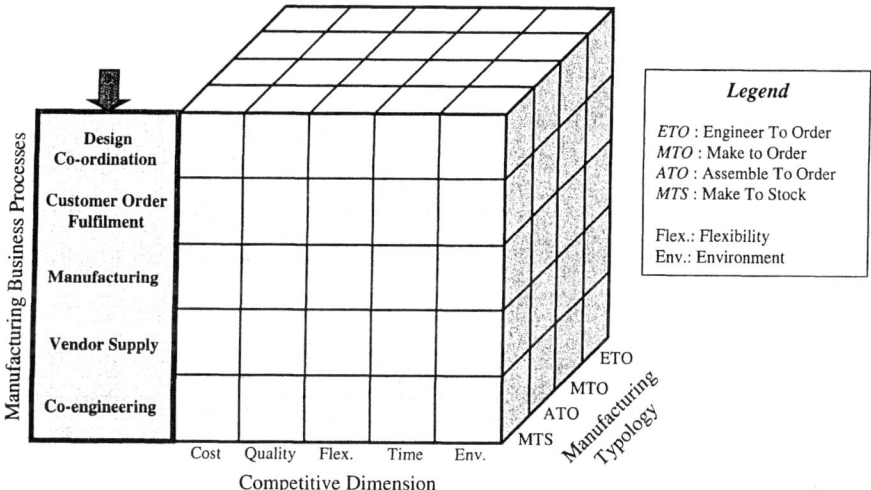

Figure 4.6 The AMBIT performance measurement framework

- *Customer Order Fulfilment* process refers to all of the activities which directly relate to the planning, data transmission, control and co-ordination of customer requirements within the manufacturing process and the delivery of the product to the customer. The goal of this process is to deliver high quality products at competitive prices, which satisfy customer requirements in the shortest time possible.
- *Vendor Supply* process concerns all of the activities involved in the co-ordination of supplier capabilities. It is also involved with the planning requirements and the delivery of these requirements to the manufacturing process. The goal of this process is to supply high quality components etc. to the manufacturing process within the shortest time possible and at the keenest prices.
- *Manufacturing* process accounts for all of the activities directly involved in either the physical production of the product (e.g. machining systems) or the delivery of the organisational service. The goal of the manufacturing process is to manufacture high quality products in the shortest time possible while managing manufacturing costs at an acceptably low level.

- *Design Co-ordination* process is concerned with all of the activities directly involved in the design and development of a product, subject to customer specifications and its release to manufacturing. The design co-ordination process thus includes the design of the manufacturing process. Its goal is the design of a high quality product, which satisfies both customer requirements and manufacturing process requirements, in the shortest time possible.
- *Co-engineering* process refers to all of the activities directly involved with the co-ordination of supplier capabilities into the product design process. The objective of the co-engineering process is to effectively use the skills and experience of suppliers in designing a high quality product which satisfies both the requirements of the customer and the manufacturing process.

Each of these processes can be further decomposed in order to show the sequence of activities occurring between each main element in greater detail. As an example, Figure 4.7 shows the partial decomposition of the customer order fulfilment process within a *make-to-stock* environment.

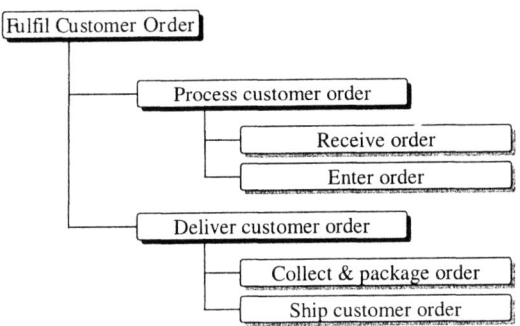

Figure 4.7 Decomposition of the customer order fulfilment (COF) process within a make-to-stock environment

In a *make-to-stock* environment (see section 4.3.3), the key processes include the forecasting of the manufacturing requirements, the manufacturing of the products, the maintenance of the stocks and the fulfilment of the customer orders. From the fulfil customer order perspective, two key processes are:
- process customer order, and
- deliver customer order.

All of the organisational elements and their interactions through the macro business processes, impact the overall objectives of the organisation in terms of its financial and market position which in turn are set by, and should satisfy, the organisation's corporate strategy. These processes represent the first axis of the AMBIT performance measurement framework. The next step concerns the identification of the macro measures of performance, which comprise the second axis of the AMBIT performance measurement framework.

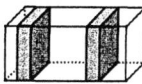

4.3.2   Macro Measures of Competitive Performance Axis

The second axis of the performance measurement framework (Figure 4.8) is defined by a set of macro measures of performance, which are used to identify dimensions along which the organisation must compete in the marketplace.

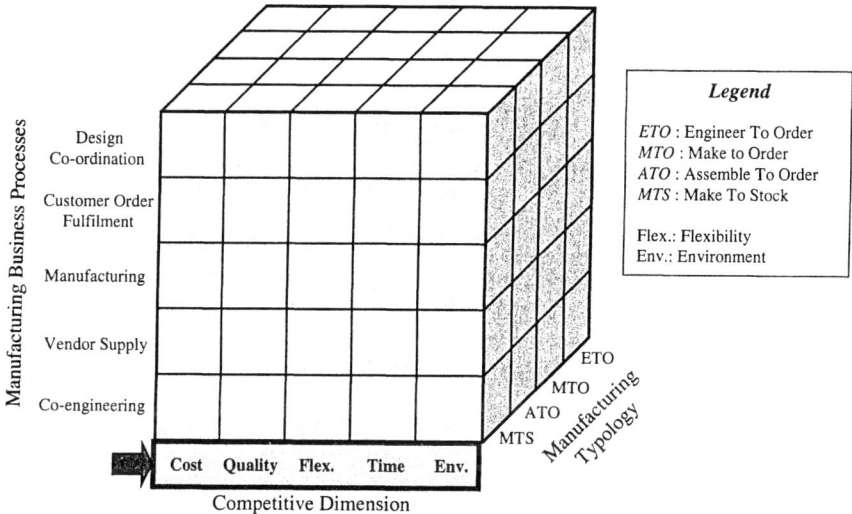

Figure 4.8 The AMBIT performance measurement framework

Up until the 1970s, price was generally regarded as the basis on which an organisation competed. However, recent changes in the manufacturing environment have shifted the focus from cost to alternative competitive measures (around which an organisation must structure its competitive strategy). Whilst there is no standard competitive measure within a given industry, an organisation can attach different emphases to different competitive dimensions. The predominant product and process strategies used by organisations, in maintaining a competitive advantage from the 1970s to the 2000s are shown in Table 4.2.

Table 4.2 Product and process strategies [after Brown 1996]

	1970s	1980s	1990s	2000s
Market Characteristics	Stable	New competition	Increased competition	Volatile, Dynamic
Market Requirements	Price	Price, Quality	Price, Quality	Price, Quality, Uniqueness
Product Strategy	Few varieties	Quality	Quality, Wide range, Differentiation	Differentiation, Constant flow of new products

	1970s	1980s	1990s	2000s
Process Strategy	Efficiency	Efficiency, Quality	Efficiency, Quality, Flexibility	Efficiency, Quality, Flexibility, Innovation
Manufacturing's Role	Long runs	Quality	Short runs	Strategic, Ability to adapt quickly

Within the AMBIT manufacturing model, the macro measures of competitive performance are as (Figure 4.9) Cost, Quality, Flexibility, Time and Environment.

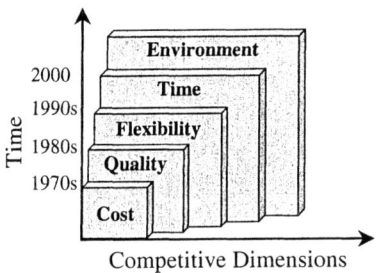

Figure 4.9 Trends in competitive dimensions

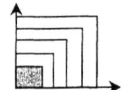 *Cost as the macro measure of competitive performance*

Cost management is a key element for survival in a highly competitive environment. The active search for opportunities to decrease the total cost of a product has traditionally focused on reducing costs of existing products.
– Everaert *et al.* 2002

Cost historically represented the major basis of organisational competition up until the 1970s. During that period, cost leaders pursued organisational cost reductions in order to support competitive advantage. An organisation traditionally reflecting a low-cost position possessed a favourable defence against:

- **Competitors.** By producing its products at very competitive costs, an organisation could compel its competitors to compete away their profits through rivalry.
- **Suppliers.** By providing greater flexibility in dealing with input cost increases, supplier power was reduced.
- **Buyers.** Buyers would use their power to drive price reductions until they reflected those of the most cost-effective supplier.

During the post 1970s, quality emerged as an important issue in the competitive scheme of things.

Quality as the macro measure of competitive performance

Quality can mean many things. Thus, when trying to define exactly what quality is, it can be useful to look at it from a number of different perspectives i.e. [Tamimi *et al.* 2002]:

- Philosophical view – quality means excellence.
- Product based view – quality is a measurable characteristic of the product.
- Value based view – quality refers to conformance at an acceptable cost or performance at an acceptable price.
- Manufacturing based view – quality refers to conformance to specifications.

Meanwhile, Hewlett-Packard (HP) defines quality as "a set of product or process attributes that provide value to customers that meets or exceeds their expectations. The organisation recognises two kinds of customers: external and internal. External customers are companies and individuals who purchase HP goods and services. Internal customers are people and departments inside Hewlett Packard whose performance is directly affected by the performance of co-workers" [Hewlett-Packard 1990]. In the years from 1970 to 1980, quality emerged as the major distinguishable feature in manufacturing performance between Western and Japanese manufacturing organisations. In today's terms it is vital that organisations view quality as an holistic organisational activity if they are to gain a competitive advantage. Holistic refers to flexibility, speed, cost (especially in the reduction of waste), delivery capabilities and service etc. Thus, it is important that organisations recognise the inherent strategic importance of quality, for organisations that compete "more on quality and less on price are less vulnerable to competition form low cost producers and from adverse movements in the exchange rate" [DTI 2002]

In response to the increasing number and capabilities of new entrants and competitors into global markets and the increasing diversity of product choice provided to the customer, quality has emerged as an important strategic weapon. If an organisation has a strongly established total quality mindset, it is "dedicated to the customer's satisfaction in every way possible ... all activities of all functions are designed and carried out so that all requirements of the ultimate customer are met and expectations exceeded" [Ciampa 1992]. The maintenance of a continuous quality programme, for the purpose of sustaining and enhancing quality in the organisation, is representative of the Japanese principle of Kaizen. The essence of kaizen is simple. It means improvement, continuous improvement involving everyone within the organisation. The underlying principle of kaizen maintains that our way of life, be it our working life, our social life, or our home life, deserves to be constantly improved.

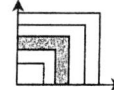 *Flexibility as the macro measure of competitive performance*

In adapting to an environment of continuous change, it is important that an organisation be flexible in terms of volumes and product varieties. Such flexibility improves an organisation's ability to manufacture the diverse range, mix and volume of products demanded by today's discerning customer. Flexibility supports the:

- Accelerated response to volume and product mix changes.
- Rapid customisation of the product in order to satisfy customer demands.
- Reduced time to market for new products and product variants.

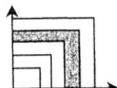 *Time as the macro measure of competitive performance*

Today's markets are increasingly time-based competitive. Time-based competition has altered the organisational focus from the creation and defense of a competitive position to one that is based on speed and flexibility. Managing time has enabled successful organisations not only to reduce their costs but also to "offer broad product lines, cover more market segments and upgrade the technological sophistication of their products" [Stalk *et al.* 1993]. Manufacturers who are the first to market with new products and who provide for the rapid delivery of products to their customers possess a competitive advantage over their competitors with respect to higher levels of customer satisfaction, greater profitability and increased market share [Chen *et al.* 2001].

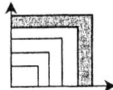 *Environment as the macro measure of competitive performance*

> Societal demands and the significant international environmental legislation that has forced companies to undertake and participate in extensive environmental activities. Commensurate with the considerable growth in environmental regulation, organisations increasingly recognize the importance of addressing environmental issues effectively. – Dunk 2002

As manufacturing enters the 21st century, there is a growing public awareness of the environmental impact of manufacturing processes and products. Whilst society is exerting pressure on manufacturers to create environmentally benign products; legislation is also becoming increasingly strict with respect to product and process design, energy utilisation and the use of recyclable materials. The eco-tax represents governments' response to the increasing demand for environmental legislation. The aim of eco-taxes, which are also known as green taxes, is simple. The aim is to encourage a greater awareness of the irreparable environmental damage that can be caused due to heedlessness and greed. Eco-taxes make people pay for their use of natural resources e.g. the tax on motor fuel might be raised in order to encourage people to use more energy-efficient cars or to use cars less frequently, or indeed both. Eco-taxes work on the principle of shifting the balance from the use of natural resources (which are currently over-used) towards the use of human resources (which are currently under-used). Regarded as a wider restructuring of taxation, Robertson suggests that environmental taxation will encourage not just environmentally benign development but also enhance economic performance, higher employment levels, a cleaner and an intensified awareness of the environment and greater economic justice within and between nations [Robertson 1997].

Lawrence *et al.* suggest that an increasing numbers of world-wide organisations are achieving certification to the environmental management system (EMS) standard ISO 14001. Some of the advantages accruing to an organisation from an EMS implementation include [Lawrence *et al.* 2002]:

- a commitment to the safeguard of the environment,
- cost savings and improved management control,
- satisfying customer expectations,
- enhanced environmental performance,
- being at the forefront of legislation, and
- increased employee motivation.

Clearly, organisations which comply with environmental regulations and provide environmentally benign products, which can eventually be dis-assembled or reused, will gain a significant competitive advantage over their competitors. Indeed, organisational compliance with environmental legislation contributes "to national economic performance, environmental and social well-being and international economic justice" [Robertson 1997]. Also, as already outlined, the AMBIT manufacturing model, based on the extended enterprise model, caters to the recovery of resources and end of life products.

Kurakawa *et al.* describe an internet based tool, called the Green Browser, which supports designers "to clarify trade-off relationships among the strategic goals over the life-cycle and to share environmental information with customers" [Kurakawa *et al.* 1997]. A conceptual scheme of the Green Browser is called the Green Life-Cycle Model. It consists of the strategy model, the process model and the object model, which are linked together. "By using the Green browser, designers can build the green life-cycle model and determine the relationship of a design requirement with others, so the strategy of the product is clarified" [Kurakawa *et al.* 1997].

The macro measures of competitive performance: time, cost, quality, flexibility and the environment are considered valid for the majority of manufacturing organisations. They are valid, because the goal for such organisations is generally for a highly flexible, environmentally benign process, which facilitates the manufacture and speedy delivery of a diverse range of high-quality, cost-effective, environmentally benign products. It is therefore important that these competitive dimensions be given due consideration in the determination of any organisational strategy. Each of the five macro business processes: customer order fulfilment, vendor supply, manufacturing, design co-ordination and co-engineering can be measured along one or more of these macro measures of competitive performance.

The next step concerns the identification of the manufacturing typology, which comprises the third axis of the AMBIT performance measurement framework.

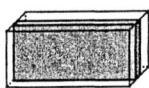

4.3.3 Manufacturing Typology Axis

The third axis of the AMBIT performance measurement framework is defined by the type of manufacturing system under study (Figure 4.10). Manufacturing systems are classified on the basis of common characteristics, in order to further the study and understanding of such systems. Whilst a number of classification approaches exist, a selection of such approaches is presented in Table 4.3. Although we will discuss all three possible approaches outlined in Table 4.3, we have elected to use the customer order decoupling point classification system in our AMBIT framework, because we feel that it is most appropriate of the three classification systems in today's market driven environment. Of course we could equally have used either of the other two classification systems.

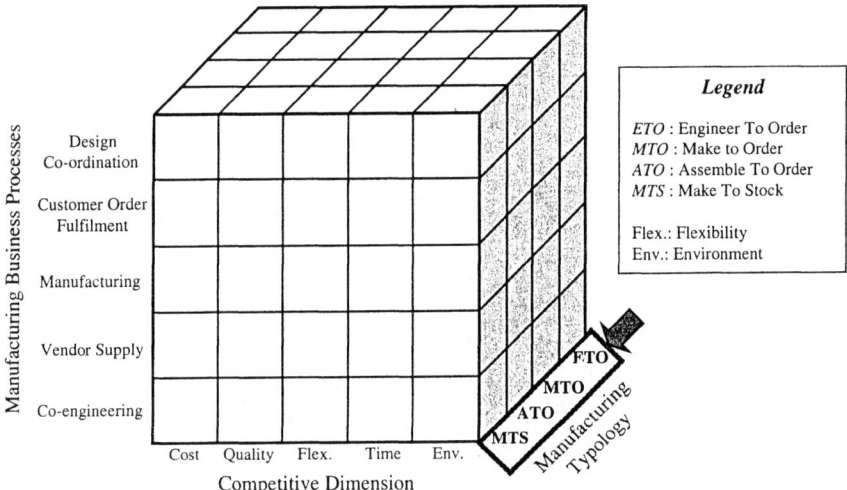

Figure 4.10 The AMBIT performance measurement framework

Table 4.3 Classification of manufacturing systems and their associated typologies

Classification Approach	Typology
Industry Sector	Capital goods, Fashion goods, Durables and Commodities
Volume/Variety	Jobbing shop, Batch, Repetitive and Mass production
Customer Order Decoupling Point	MTS, ATO, MTO and ETO

The manufacturing system classification approaches, depicted in Table 4.3, use different foundations on which the classification is based. As an example, the customer order decoupling point approach describes a classification approach based on the type of relationship between the customer and the manufacturing organisation. Meanwhile, the volume/variety approach describes a classification approach based on the degree of variety across the product range and the relative volumes of each product variety produced.

Classification by Industry Sector

The Uncertainty-Complexity grid as defined by Puttick, assigns manufacturing systems to one of four business sectors. This assignment is based on the complexity of the product being produced and the uncertainty of the market which it is seeking to address [Puttick 1986]. Complexity measures the number of product items and the product or process knowledge entities which must be managed. Uncertainty measures both the product range demanded by the customer, and the uncertain nature of the marketplace. Based on the Uncertainty-Complexity grid, four different manufacturing sectors can be identified: capital goods, fashion goods, commodities and durables (Figure 4.11). From a designer's perspective, the uncertainty-complexity grid is a tool, which supports the identification of the most beneficial elements of process design, product development and environmental issues being included in the final design [DTI 1993].

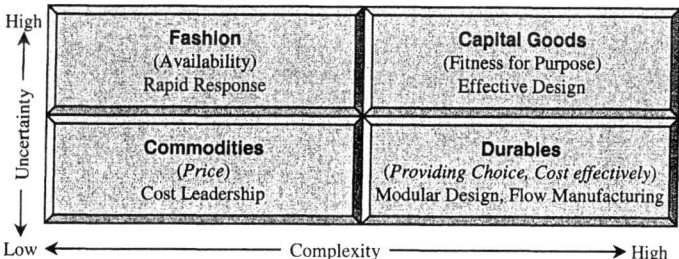

Figure 4.11 The uncertainty-complexity grid [after Puttick 1986]

 *Fashion Industry*

Fashion manufacture generally refers to the manufacture of clothing, crafts, leisure goods, entertainment etc. The demand in these manufacturing sectors is normally very uncertain while the products are normally quite simple. Manufacturing within this industrial sector is typified as follows:

- An emphasis on the availability of goods.
- A focus on rapid response by means of:
  o integration between design and manufacturing,
  o tight links between factory, suppliers and customers, and
  o surplus supply control.
- Simple fashion products or parts for repair.
- Flexible job-shop based manufacturing.
- A wide range of customers.
- Variable supplier base.
- Close linkage between customers and the factory.
- The need for integration between design and manufacturing.

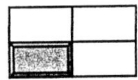 *Commodities Industry*

Commodities manufacture generally refers to the manufacture of everyday products such as food, paper and other simple products with relatively predictable demand. Manufacturing within the commodities industrial sector is characterised by:

- A focus on price and cost leadership.
- A wide range of customers.
- The provision of a standard range of simple products.
- "V" type of plant configuration: manufacture of a high number of product derivatives from a narrow range of raw materials.
- Product packaging (and possible varieties) can play a significant role.
  o Continuous manufacturing on process/transfer lines.
- Manufacturing occurs within a make-to-stock (MTS) environment.
- The need for optimum capacity utilisation via optimum production sequences.
- Strategic sourcing for key materials.
- Purchasing:
  o Few major suppliers.
  o Strategic sourcing for key ingredients.
  o Vendor scheduling or call offs
- High volume of 'simple' orders.
- Efficient distribution network, including finished goods inventory planning and control.

 *Durables Industry*

Durables manufacture generally refers to the manufacture of complex goods in a more certain market (e.g. consumer electronics, computers and automotive manufacturing). The key characteristics of durables industries are:

- An emphasis on the provision of a quick and cost-effective choice.
- A customer base which covers a range of medium and major customers.
- "T" type of plant configuration: High volume of sub-assembly orders merged into medium volume of customised assemblies.
- The production of a standard range of products with limited options.
- The assembly of products, driven by Final Assembly Schedules (FAS), to customer orders using sub-assemblies from stock, driven by Master Production schedule (MPS).
- Modular design flow-manufacturing.
- Range of medium and major customers with high volume of customer orders.
- Significant distribution chains.
- Purchasing is a low sourcing activity, with preferred suppliers.
- Close coordination with the suppliers.
- The provision of an after sales and spare parts service.
- Fast product development complying with environmental regulations.

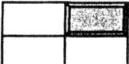 *Capital Goods Industry*

Capital goods manufacture generally refers to the manufacture of specialised equipment, which is characterised by high complexity and uncertain demand. An example is the manufacture of ships, aircraft, railways rolling stock, machine tools, etc. Manufacturing within the Capital Goods sector tends to have the following key characteristics:

- An emphasis on effective design and product fitness for use.
- A small customer base.
- A complex product base.
- Customised products.
- "A" type of plant configuration:
  o Flexible production processes.
  o Many low/medium volume works orders.
  o Many sub-assemblies/components/raw materials.
  o Single (major) final assembly.
  o Low production volumes.
- Manufacturing occurs within an engineer-to-order (ETO) environment.
- Project management principles needed to manage customer orders.
- Uncertain Demand.
- Purchasing is a high sourcing activity.
  o Typically discrete ordering.
  o Many suppliers and possible sub-contracting activities.
- Simple distribution systems.

Classification by Volume/Variety

A manufacturing organisation can also be categorised based on the degree of variety across its product range and the relative volumes of each product variety produced. Any manufacturing organisation can thus be said to embrace either a mass, repetitive, batch or jobbing shop production manufacturing approach (Figure 4.12). This classification approach refers to the discrete parts production of individual items.

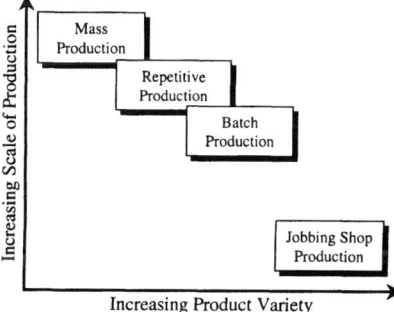

Figure 4.12 Classification of discrete production

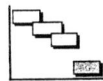 *Jobbing shop production*

Jobbing shop production is characterised by low volume production runs of non-standardised, complex and often one of a kind products (e.g. aeroplanes). To produce the diverse product range typical of a jobbing shop production system, a highly flexible production system and good sourcing capabilities are required.

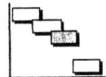 *Batch production*

Batch production is typified by the medium to high volume production of a medium range of product varieties. In batch production, a product is produced in small batches by a series of operations. Each operation is usually carried out on the entire batch, before the start of the next operation. Due to the often-unstable nature of buyer behaviour, it is important that the batch production system be reasonably flexible. .

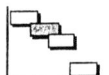 *Repetitive production*

Repetitive production refers to the fabrication, machining, assembly and testing of discrete, standard units, which are produced in volume. It also refers to the fabrication, machining, assembly and testing of products assembled in volume from standard options. A typical repetitive manufacturing system is characterised by long runs of complex products produced in lower volume quantities. In the ideal repetitive production setting, parts are directly transferred from one work centre to another. Whilst repetitive manufacturing evolved in response to the increased flexibility of automatic manufacturing, repetitive manufacturing organisations try to remove the mass produced look from their products and introduce some customisation, albeit at the later stages of the production cycle. Successful operation within this sector demands an increasing emphasis on faster product development times and improved customer service.

 Mass production

Mass production is typified by the large volume production of relatively few, highly standardised products. A manufacturing climate characterised by stable product demand and little product variety represents the ideal setting for mass producers, where interaction between the mass producer and the customer is negligible if existent at all (e.g. the 'Model T' by Ford was a prime example of a mass produced product).

It is rare that a discrete parts manufacturing system will be either pure mass, repetitive, batch or jobbing shop production. The majority of discrete parts manufacturing systems exist on a continuum between jobbing shop production and pure mass production.

Classification by Customer Order Decoupling Point

> Customer driven manufacturing, in which production activities are driven by customer orders, is the key concept for the factory of the future. This concept results from the trends of production processes of small batch sizes and customised products over the past decade.   – Yeh 2000

Today's customers are increasingly discerning. No longer are they willing to accept a low range of standardised products. The traditional approach where the manufacturer's interaction with the customer was negligible and high volume standard components were manufactured, stored in a warehouse and then withdrawn by the customer, is no longer feasible (Figure 4.13 – A). In response to intensifying pressures exerted by customers for greater participation and customised products, manufacturing organisations are becoming increasingly flexible, producing higher quality products and adopting customer-driven approaches. With the absence of a finished goods inventory buffer, from which the customer can order the product, the customer-driven approach encourages direct customer interaction with the manufacturer in the expression of his preferences concerning the product (Figure 4.13 – B).

With the movement of manufacturing towards the production of customer specified products, the relationship between the manufacturer and his suppliers and customers is vital in maintaining a competitive stance. The increasing customer demand for customised products and faster delivery times, means that manufacturers are more than ever required to establish closer links with their suppliers in order to reduce material supply lead times and improve delivery performance.

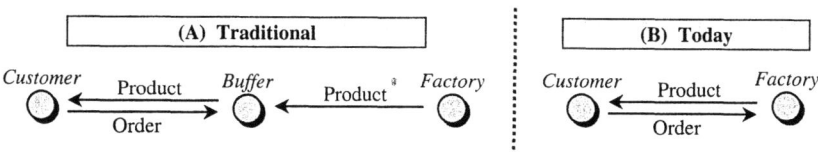

Figure 4.13 Evolution of manufacturing

Figure 4.14 shows the information and material flows between the manufacturer and his corresponding suppliers and customers.

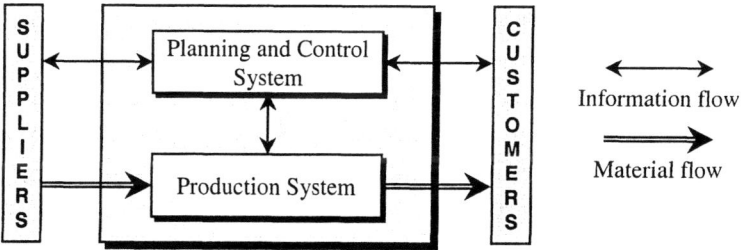

Figure 4.14 Information and material flows

The growing emphasis on customer-driven manufacturing is resulting in the progressive movement of organisations along the manufacturing system typology continuum illustrated in Figure 4.15. This movement reflects a move away from a make-to-stock environment and towards an engineer-to-order environment.

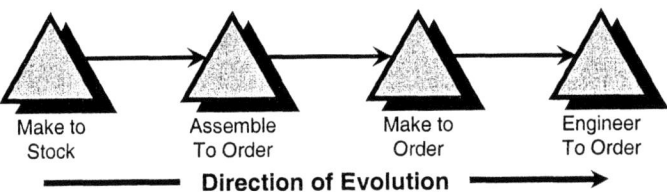

Figure 4.15 Manufacturing system typology continuum

At different points along this continuum, material is dedicated to a particular customer order. These points are called customer order decoupling points. The positioning of these points is important, for it defines the parts of the process driven by customer orders and the parts driven by forecasts. Based on the ratio between the parts of the process which are driven by customer orders and those which are driven by forecasts (i.e. the point in manufacturing at which a product is firmly committed to a particular order) a manufacturing organisation's operation can be classified as one of the following (Figure 4.16):

- Make-to-stock (MTS).
- Assemble-to-order (ATO).
- Make-to-order (MTO).
- Engineer-to-order (ETO).

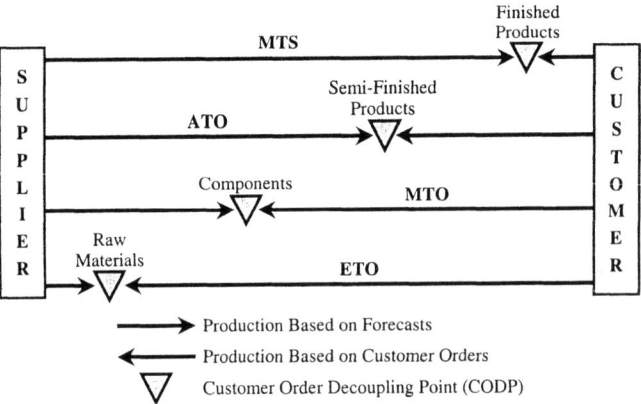

Figure 4.16 Customer order decoupling point

The manufacturing system approach adopted by an organisation influences the manner in which that organisation conducts its manufacturing, planning and control activities.

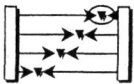 *Make-to-stock (MTS)*

MTS typifies the manufacture of products based on a familiar and relatively predictable demand mix, where product life cycles are reasonably long and stable. In an MTS organisation, a stock of finished products is maintained, from which customer orders are filled. Interaction with the customer is minimal and the production volume of each sales unit is high. Whilst an MTS system offers quick product delivery times, the disadvantages of this system include the following:

- Inventory costs are high, due to the need for a large supply of finished goods in order to act as a buffer against uncertain demand and stockouts.
- Due to the minimal interaction with customers, customers are unable to express preferences with respect to the product design.

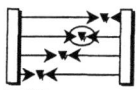 *Assemble-to-order (ATO)*

ATO organisations maintain a stock of semi-finished products, so that following receipt of an order for a particular configuration, the relevant sub-assemblies can be assembled. In the ATO environment, the same core assemblies are generally used for the majority of products. Furthermore, whilst product routing is fixed, product delivery time is relatively short and based on the availability of major subassemblies. A typical example of this type of manufacturing is that of personal computers.

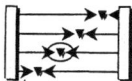 *Make-to-order (MTO)*

MTO organisations maintain a stock of standard components, so that following the receipt of an order for a particular design; the product is manufactured from these components. Whilst the product is not specified until a customer order is received, finished products from this system are partially one of a kind, but not pure one of a kind because the final product is not usually designed from a particular specification. Interaction with the customer is extensive and is based on sales and engineering. Product delivery times range from medium to long, whilst promises for completion of orders are based on the available capacity in manufacturing and engineering. The manufacture of machine tools and many capital goods are examples of this type of MTO manufacturing.

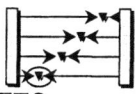 *Engineer-to-order (ETO)*

ETO represents an extension of the MTO system, with the major difference being that the engineering design of the product is almost totally based on customer specifications. Customer interaction is greater and true one of a kind products are engineered to order.

The comparative characteristics of these manufacturing systems are presented in tabular form in Tables 4.4 and 4.5.

Table 4.4 Customer interaction in different manufacturing systems

Aspect	MTS	ATO	MTO	ETO
Interface between manufacturing and customer	Low / Distant	Primarily at sales level	Engineering and sales level	Primarily at engineering level
Delivery time	Short	Medium	Variable	Variable
Production volume of each sales unit	High	Medium	Low	Very low
Product range	Low	Medium	High	High
Order promising (based on)	Available finished goods inventory	Availability of components and major subassemblies	Capacity,(m anufacture, engineering)	Capacity (manufacture, engineering)

Table 4.5 Production planning in different manufacturing systems

Aspect	MTS	ATO	MTO	ETO
Basis for production planning and scheduling	Forecast	Forecast and Backlog	Backlog and Orders	Customer Orders
Handling of demand and uncertainty	Safety stocks of sales units	Over-planning of components and sub-assemblies	Considerable uncertainty exists	No control

The selection of a particular manufacturing strategy modifies the structure of each of the macro business processes. For example, in a make-to-stock (MTS) system, order processing (which is an element of the customer order fulfilment business process) is not regarded as very important, whilst in an engineer-to-order (ETO) system, order processing is regarded as very important.

4.4 Generate Strategic Performance Indicators (SPIs)

At this stage of our discussion of the AMBIT approach, the business goals - defined as critical success factors (CSFs), have been mapped onto the AMBIT performance measurement framework in terms of the three dimensions already outlined. Now the SPIs need to be generated (Figure 4.17).

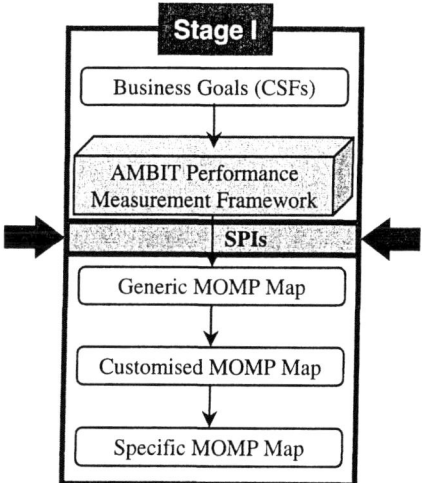

Figure 4.17 Progress in the AMBIT approach - Stage I

The example in Figure 4.18 illustrates the translation of critical success factors (CSFs) into corresponding strategic performance indicators (SPIs). In this example, the CSFs are:
- Shorten order delivery time.
- Rapid order confirmation.
- Increase delivery accuracy.

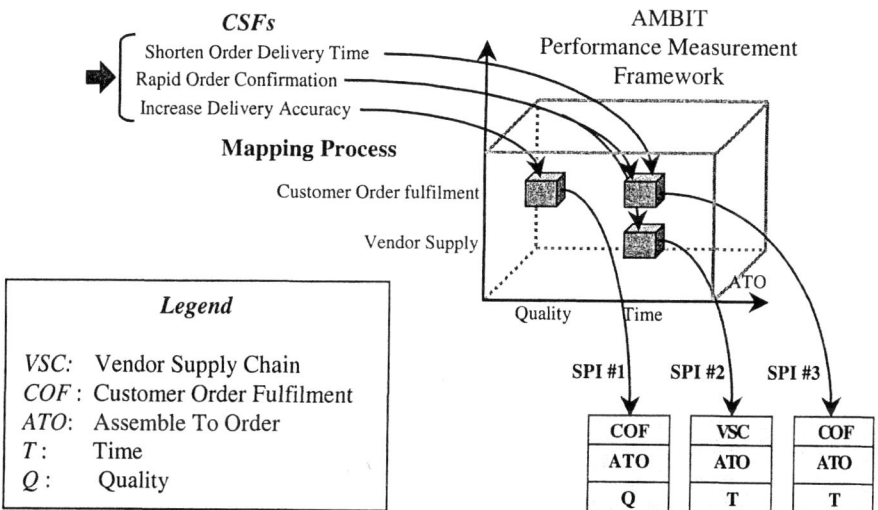

Figure 4.18 Using the AMBIT performance measurement framework

As it is difficult to relate these high level CSFs to the strategy pursued by manufacturing, they are expressed in the form of strategic performance indicators (SPIs). In the identification of the SPIs, the relevant manufacturing typology, the associated manufacturing business processes and the corresponding competitive dimensions (i.e. macro measures of performance) are first selected. The macro measures of performance are quality (Q) and time (T). Given the nature of the CSFs, the corresponding manufacturing business processes are:

- Shorten order delivery time: The customer order fulfilment (COF) process.
- Rapid order confirmation: The vendor supply chain (VSC) process and the customer order fulfilment (COF) process.
- Increase delivery accuracy: The customer order fulfilment (COF) process.

In this example, the organisation maintains a stock of semi-finished products, so that following the receipt of an order for a particular configuration, the relevant sub-assemblies can be assembled. Thus, the organisation is operating in an assemble-to-order environment (ATO). Taking all of this information into account, the SPIs relating to the CSFs are:

- Shorten order delivery time: Time (T) in the customer order fulfilment (COF) process in an ATO environment – (COF/ATO/T).
- Rapid order confirmation: Time (T) in the vendor supply chain (VSC) process in an ATO environment – (VSC/ATO/T) and Time (T) in the customer order fulfilment (COF) process in an ATO environment – (COF/ATO/T).
- Increase delivery accuracy: Quality (Q) in the customer order fulfilment (COF) process in an ATO environment – (COF/ATO/Q).

As shown in Figure 4.18, each mini-cube represents an SPI. However, one SPI could be related to more than one CSF. For example, the following CSFs: Shorten Order Delivery Time and Rapid Order Confirmation both affect SPI#3.

SPIs basically refer to the indicators which are useful in measuring the performance of an organisation from a strategic perspective. The mapping of the CSFs (which represent the strategy of the organisation) onto the AMBIT performance measurement framework results in a set of SPIs. These SPIs in turn can be used to identify whether or not the organisation is satisfying its mission objectives, by measuring the performance of the individual manufacturing business processes in terms which are appropriate to the CSFs of that organisation.

4.5   Develop the Generic MOMP Map

The set of strategic performance indicators (SPIs) which collectively represents the motivation of the organisation, based on its business goals are further decomposed into a map of generic measures of manufacturing performance (MOMPs) which represent operational level measures of performance (Figure 4.19). A MOMP therefore represents the association of the manufacturing business processes with the corresponding competitive dimensions. Thus, an SPI can also be considered a high-level MOMP. A sample list of performance measures is presented in the Appendix to this book.

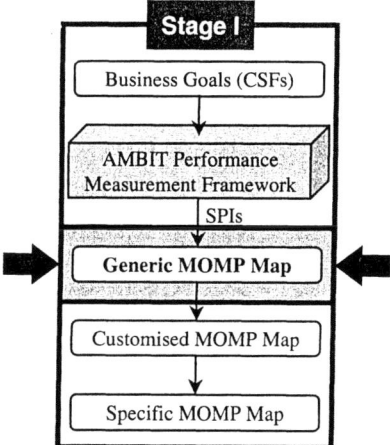

Figure 4.19 Progress in the AMBIT approach - Stage I

As already stated, the high-level nature of the strategic performance indicators (SPIs) makes quantification difficult. To overcome this, the SPIs are decomposed into measures of manufacturing performance (MOMPs). This decomposition of the SPIs into MOMPs depicts a movement from a high level perspective to a lower level perspective. In the AMBIT approach, the MOMPs aggregate upwards to the SPIs (Figure 4.20).

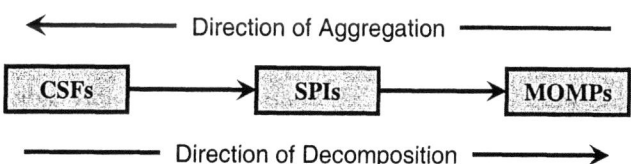

Figure 4.20 CSFs are decomposed into SPIs, which are in turn decomposed into MOMPs

The AMBIT performance measurement framework provides a structured approach to the identification of these MOMPs. As we will see later in this chapter, these MOMPs will be quantified for each sub-process, sub-sub-process (Figure 4.21), and so on until the level of detail appropriate to the organisation under study is achieved.

As an example, the approach adopted in the derivation of the MOMP templates and the definition of associated MOMPs for the customer order fulfilment (COF) process is described as follows

Step 1 - Identify the associated sub-processes.

Step 2 - Determine the impact of the four manufacturing typologies on the process and its associated sub-processes.

Step 3 - Identify the appropriate measures from the sub-processes, in terms of the macro measures of competitive performance i.e. cost, quality, flexibility, time and environment.

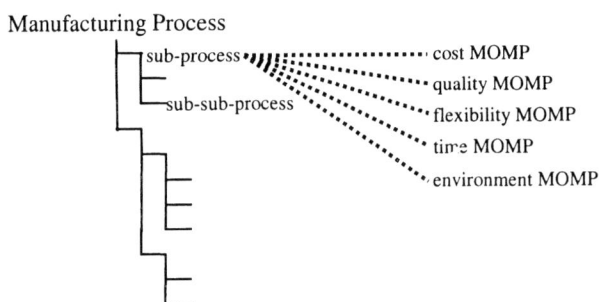

Figure 4.21 A manufacturing process having five MOMPs for a sub-process

4.5.1   Step 1: Identify the associated sub-processes

As previously outlined, the customer order fulfilment (COF) process includes the activities directly associated with the planning, control and co-ordination of customer requirements with the manufacturing process, such that the specified product is delivered to the customer, in the correct quantity and within the designated timeframe. The elements associated with this process are shown in Figure 4.22. This is of course a highly simplified representation, where the distribution activity is ignored.

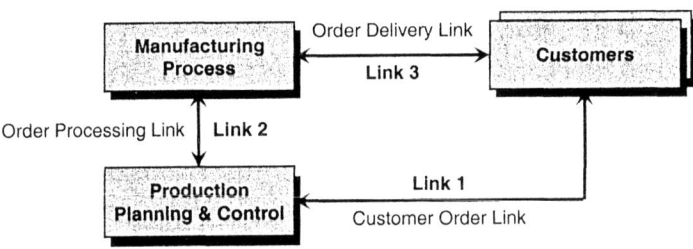

Figure 4.22 Key links in the customer order fulfilment process

The links in the COF process, described by the links numbered, 1, 2 and 3 respectively, are as follows:

- **Link 1** represents the customer order link. This link is used to communicate the customer orders to the organisation or the process. It describes the initiation and maintenance of customer contact with the order fulfilment process. The types of flows occurring in this link include information requests, orders, order negotiations, etc. Link 1 is influenced by the manufacturing approach of the organisation (i.e. assemble-to-order (ATO), engineer-to-order (ETO), make-to-stock (MTS) and make-to-order (MTO)). In an MTS organisation, the customer order link deals with the processing of large volumes of orders for 'standard' products. Meanwhile in an ETO organisation, the customer order link handles a

small number of very complex and unique orders and requires the detailed involvement of technical and design support staff as well as sales staff
- **Link 2** is the manufacture customer order link. It is concerned with processing the customer orders and ensuring that everything is readily available for the manufacturing processes in order to fulfil these orders. The types of flows occurring in this link include information on the schedules of all the currently *active* orders, material availability, resource availability, set-ups required etc.
- **Link 3** is the order delivery link. Whilst this link is associated with the delivery of the orders to the customers, the primary flows that occur within it include the flow of the products to the customers and the flow of information concerning the delivery of the products.

Using these links as a basis (Figure 4.22), the customer order fulfilment process, termed 'Fulfil Customer Order', can be decomposed into three lower level sub-processes (Figure 4.23 -A) i.e. process customer order (link 1), manufacture customer order (link 2) and deliver customer order (link 3). It should be borne in mind that this illustration is a very simple high-level breakdown of the customer order fulfilment process that is independent of the manufacturing typology.

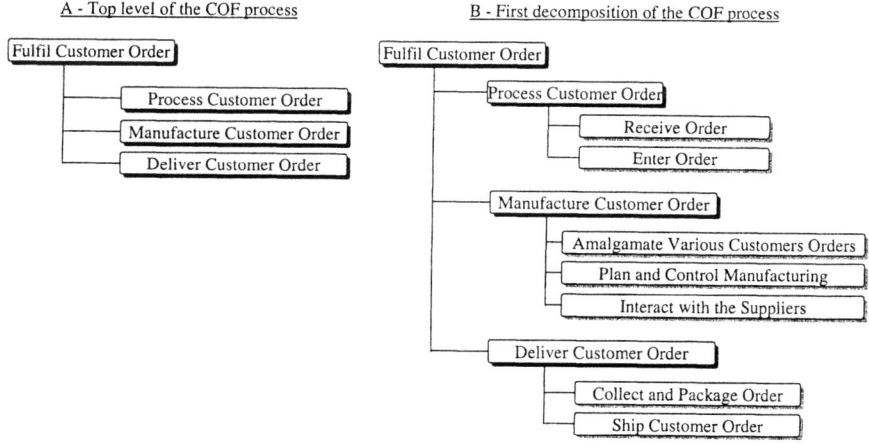

Figure 4.23 A simplified customer order fulfilment (COF) process (independent of the type of manufacturing system).

Each of these sub-processes can in turn be further decomposed (Figure 4.23-B):
- Process customer order could have two sub processes:
  a) Receive order: refers to all of the activities related to interacting with the customer and ensuring that the placed order is either compatible with the product line or is manufacturable.
  b) Enter order: refers to the integration of the confirmed customer order with the organisation's information system, so that other information sub-systems can access and extract pertinent information.
- Manufacture customer order could have three sub processes:
  a) Amalgamate various customer orders: refers to the decomposition of the individual customer orders into specific manufacturing requirements and

the quantification of the net requirements for each product across all customers. For example, customer 1 requires 5 units of product A and 19 units of product B. Customer 2 requires 11 units of product A only. The net requirements for each product is 16 units of product A and 19 units of product B.

b) Plan and control manufacturing: takes care of net manufacturing requirements and all the activities associated with manufacturing, such as, planning, scheduling and the actual manufacture of the products. In practice the tasks outlined in a and b here are normally executed within the requirements planning and Bill Of Materials (BOM) processing modules of an ERP (Enterprise Resource Planning) system.

c) Interact with the suppliers: entails communicating with the materials suppliers and ensuring that the necessary raw materials are of the specified quality and quantity and are delivered on time.

- Deliver customer order could have two sub processes:

a) Collect and package order: refers to the collection of the completed order, the confirmation that it meets all the specifications given during the customer order process phase and the selection of appropriate packaging for the purpose of safe delivery.

b) Ship customer order: refers to the interaction with the delivery and transportation companies to initiate the delivery of the packaged order to the customer.

In the description of the various business processes, it is important that a specific syntax be maintained for the purpose of clarity and to facilitate the process decomposition (Table 4.6). In this context, a hierarchical decomposition approach such as IDEF0 has been found to be very useful [Lo *et al.* 2001]. IDEF0 provides a rigorous framework with a formal, easy to use syntax. It allows the user to decompose the processes and associated measures of performance to the level of granularity appropriate to his purpose.

Table 4.6 Process syntax

Process Information	Syntax
Process Name:	A unique verb or verbal phrase.
Description:	1 - 7 lines describing the process and where the process starts and ends.
Input(s):	The input which 'starts off' the process.
Output(s):	The output from the process.
MOMP:	The MOMP name(s) used to measure this process.

Figure 4.23 is a very high level and simplified view of the process that is independent of the manufacturing typology. We now develop variants of this 'model' to consider the differences that arise when we consider the characteristics of the various types of manufacturing system (i.e. MTS, ATO, MTO or ETO).

4.5.2 Step 2: Define the business processes in terms of the needs of the various styles of manufacturing system.

As the customer order fulfilment (COF) process is further decomposed, the peculiarities of particular styles of manufacturing systems or the characteristics of particular categories of manufacturing system influence the 'lower level' or detailed process steps in different ways. Figure 4.24 illustrates in a greatly simplified way, the manner in which the four categories of manufacturing system based on the customer ordering order decoupling point typology affect the first decomposition of the sub processes of the business process Fulfil Customer Order.

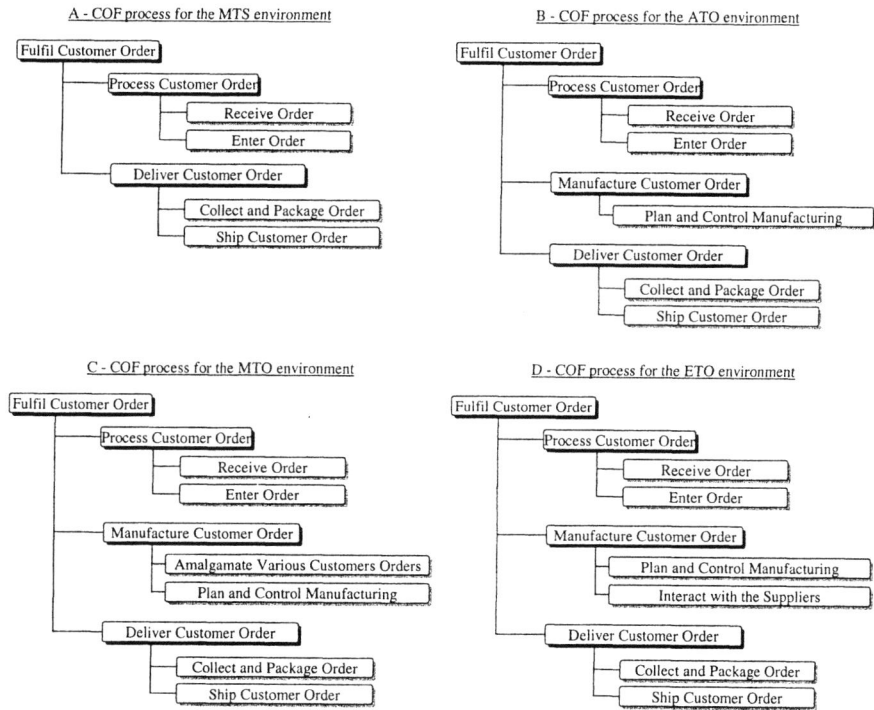

Figure 4.24 Influence of the manufacturing system on the fulfil customer order process

As shown in Figure 4.24a, manufacture customer order is not applicable in an MTS (make to stock) style of organisation, because MTS manufacturers maintain a stock of finished products ready to be shipped to the customers. Therefore, the fulfil customer order process does not have to interact explicitly with the manufacturing customer order process to satisfy any incoming customer orders. This is not to say that the manufacture customer process is not important in an MTS environment. All we are suggesting is that the manufacturing process is not directly linked to the fulfil customer order process. Meanwhile, manufacture customer order is very much applicable in an ETO and MTO (and to a lesser extent an ATO) type of organisation (Figures 4.24 d, c and b respectively). In an ETO type of organisation, the design of the product is to a great extent unique and based on the individual customer's requirements. In an MTO type of organisation, the design of the product is based on the customer unique requirements to a much lesser degree. The manufacturing function maintains a stock of standard parts and raw materials and upon the receipt of customer order, it produces the finished product according to the order specifications. Similarly, in an ATO type of organisation, manufacturing maintains a stock of semi-finished products or assemblies and upon receipt of the customer order, it assembles the finished product from standard sub-assemblies from stock, according to the order specifications.

4.5.3 Step 3: Measure the processes

The approach used in measuring the processes (in this example the 'Fulfil Customer Order' process) is as follows:
1.   All performance measures, which are used to measure processes at lower levels in the process hierarchy must be directly related to a higher-level performance measurement (Figure 4.25).
2.   The minimum number of performance measures used in the measurement of a process is limited to one whilst the maximum number is five (cost, quality, flexibility, time and environment).

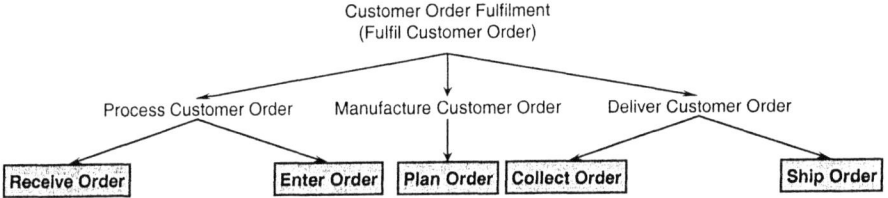

Figure 4.25 Derivation of generic MOMP map within an ATO type of organisation

The generic MOMP map reflects a partial model consisting of one or more "mini-cubes" (see Figure 4.26), each of which is defined by a single macro measure of performance, a single business process or chain and a manufacturing typology. Thus, each "mini-cube" consists of one SPI and its associated MOMP hierarchy. The generic MOMP Map comprises a number of SPI/MOMP hierarchies, normalised to remove redundancy and combined into a single map consisting of a number SPIs (one for each "mini-cube") and their associated MOMPs (Figure 4.26). Based on the CSF to SPI association and its mapping on to the generic MOMPS, in Figure 4.26, the corresponding relationship is tabulated in Table 4.7. Table 4.7 also gives explanation of the MOMPs listed in Figure 4.26.

In Figure 4.26, the Critical Success Factors (CSFs) 'Shorten Order Delivery Time' and 'Increase Delivery Accuracy' correspond to the Customer Order Fulfilment process. The macro measures of competitive performance are quality and time. The manufacturing typology is assemble-to-order (ATO), because the organisation maintains a stock of semi-finished products, so that following the receipt of an order for a particular configuration, the relevant sub-assemblies can be assembled and the final product configured to order. The corresponding strategic performance indicators will be:

▪ Quality in the customer order fulfilment process in an assemble-to-order environment (COF/ATO/Q).

▪ Time in the customer order fulfilment process in an assemble-to-order environment (COF/ATO/T).

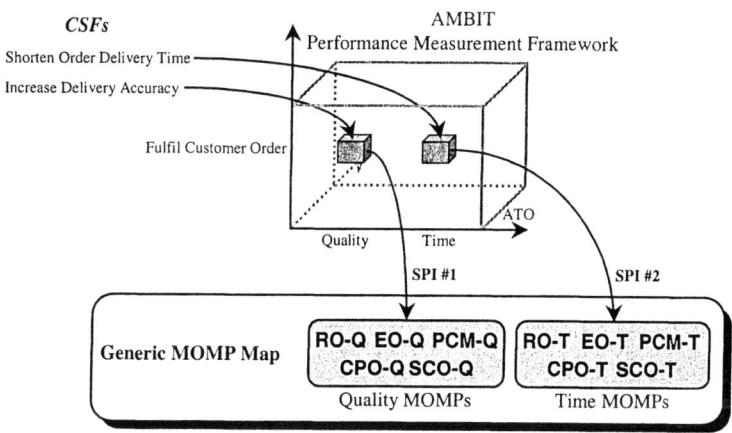

Figure 4.26 Derivation of generic MOMP map within an ATO type of organisation

Table 4.7 Generic MOMP map

SPI	MOMP abbreviation	MOMP Title
COF/ATO/T	RO-Q	Quality of the Received Order.
	EO-Q	Quality of the Entered Order.
	PCM-Q	Quality of Plan and Control Manufacturing procedures.
	CPO-Q	Quality of the Collect and Package Order.
	SCO-Q	Quality of the Shipped Customer Order.
COF/ATO/Q	RO-T	Time taken to Receive Order.
	EO-T	Time taken to Enter Order.
	PCM-T	Time taken to Plan and Control Manufacturing.
	CPO-T	Time taken to Collect and Package Order.
	SCO-T	Time taken to Ship the Customer Order.

These strategic performance indicators (SPIs) are then decomposed into associated measures of manufacturing performance (MOMPs). As described already, all of the measures used to measure a business process can be grouped under one or more macro measures of competitive performance (e.g. time, cost, quality, flexibility and the environment). If the macro measures of performance for the customer order fulfilment process are time and quality, then all of the performance measures used to measure that process should be time and quality related (Figure 4.26, Table 4.7). Thus, the AMBIT performance measurement framework provides some means of identifying a set of macro performance measures, to which all lower-level performance measures are related.

4.5.4 MOMP Syntax

As the information pertaining to a process should conform to certain syntax, so too should information pertaining to a MOMP follow a similar pattern (Table 4.8).

Table 4.8 Syntax for the MOMP information

MOMP Information	Syntax
MOMP Name:	Name given to identify the performance measure for this process.
Acronym:	Acronym for the above MOMP.
Definition:	1 - 3 lines defining the MOMP.
Description:	1 - 5 lines describing the MOMP.
Unit(s):	Measurement Unit(s) for the MOMP.
Measure(s):	List of performance measure(s) used to actually measure the MOMP.
Target Value:	A target value for the MOMP.
Target Due Date:	A target date by which to achieve the target value identified earlier.

As an example, consider the 'Total Product Development Cost' and 'Total Process Development Cost' which are cost MOMPs relating to the Design Co-ordination process, within a make-to-order (MTO) environment. The type of data which pertains to these MOMPs is shown in Tables 4.9 and 4.10 respectively. The fields labelled Target Value and Target Due Date are not filled in at this stage. They are completed when the MOMP map is particularised (in a CIM-OSA modelling sense as discussed in section 4.1 and illustrated in Figure 4.2 earlier) or customised to reflect the situation of an individual manufacturing enterprise.

Table 4.9 Syntax for the total product development cost MOMP

Information	Syntax
MOMP Name:	Total Product Development Cost
Acronym:	TPtDC
Definition:	The total cost for product development measured from product specification to the completion of a detailed product design when manufacturing commences
Description:	The total cost for product development is a cost MOMP within the Design Co-ordination process within a make-to-order (MTO) environment
Unit(s):	Cost units
Measure(s):	▪ Total product development cost ▪ Total product development cost over-run ▪ Labour cost associated with product development ▪ Investment cost associated with product development ▪ Material cost associated with product development ▪ Prototype cost associated with product development ▪ Productivity of design employees ▪ Number of design changes after the onset of production ▪ Piece cost
Target Value:	*(some value in cost units)*
Target Due Date:	*(some achievable date)*

Table 4.10 Syntax for the total process development cost MOMP

Information	Syntax
MOMP Name:	Total Process Development Cost
Acronym:	TPsDC
Definition:	The total cost for process development measured from process specification to the completion of a detailed process design when manufacturing commences
Description:	The total cost for process development is a cost MOMP within the Design Co-ordination process within a make-to-order (MTO) environment
Unit(s):	Cost units (i.e. £, $)
Measure(s):	▪ Total process development cost ▪ Total process development cost over-run ▪ Labour cost associated with process development ▪ Investment cost associated with process development ▪ Material cost associated with process development ▪ Hard tooling cost associated with process development ▪ Soft tooling cost associated with process development

Information	Syntax
Target Value:	*(some value in cost units)*
Target Due Date:	*(some achievable date)*

4.6 Customise the MOMP Map

The Customise MOMP map (or particularise in the language of CIM-OSA) stage (Figure 4.27), as already indicated, allows a MOMP map to be customised to reflect the reality of an individual manufacturing organisation and its particular business processes.

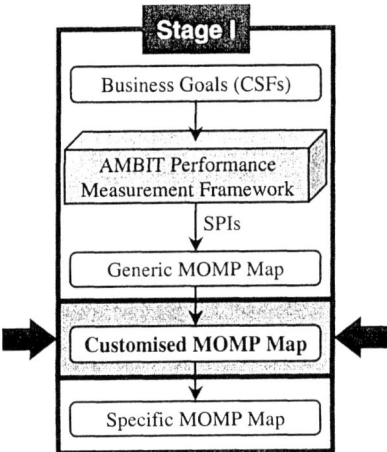

Figure 4.27 Progress in the AMBIT approach - Stage I

In a possible computer-based AMBIT tool, it is envisaged that when the user(s) enters this stage (Figure 4.28), a template of measures of manufacturing performance (MOMPs) corresponding to the selected strategic performance indicators (SPIs) for an individual business process in a defined style of manufacturing system - (e.g. ATO) - will be made available.

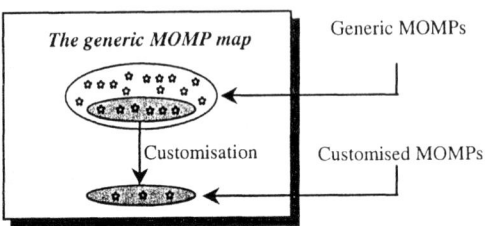

Figure 4.28 Transition from generic to customised MOMP map

Clearly this template will not reflect the details of an individual manufacturing organisation, until it is customised. Customisation of the generic MOMP map offers the user an opportunity to modify this map in order to remove performance measures that are inappropriate to the organisation under analysis. As an example, consider a hypothetical organisation, whose primary CSF is 'Shorten Order Delivery Time'. This organisation operates within an assemble to order (ATO) environment. The MOMP map which is generated based on this standard information is presented in Figure 4.29.

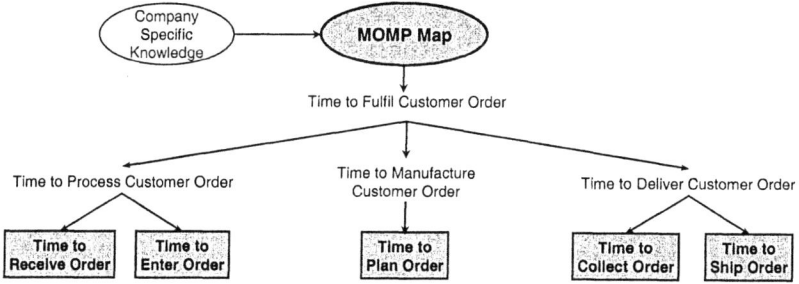

Figure 4.29 The generic MOMP map

If this organisation promotes itself under the following slogan "we ship high-quality products that you demand", then it will consequently place a high priority on getting the correct orders to the customers on time. Thus, the organisation in question will need to introduce an additional MOMP that verifies the accuracy of the customer order and that relates to a sub-process concerned with the verification of the order. This inclusion of such an additional MOMP is supported in the customisation phase, where inapplicable MOMPs are removed and additional MOMPs are introduced to reflect the reality of the organisation under analysis. This additional MOMP may be 'Time to Verify Customer Order', as depicted in Figure 4.30.

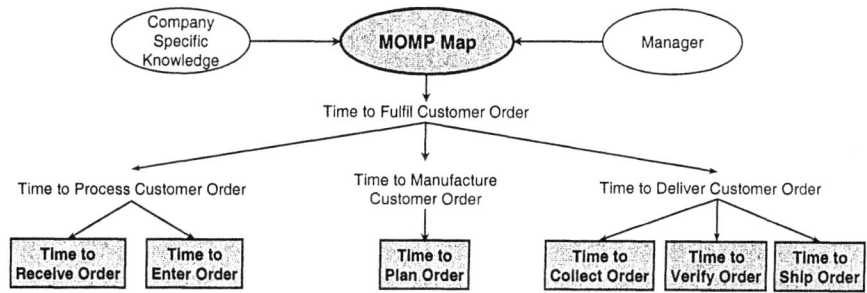

Figure 4.30 The customised MOMP map

4.7 Complete the Specific MOMP Map

The customised MOMP map must now be completed by the incorporation of organisation or company specific data. Values are assigned to the measures of manufacturing performance (MOMPs) and the strategic performance indicators (SPIs) contained in the customised MOMP Map (Figure 4.31). Specific data and targets of the type suggested in Tables 4.8 to 4.10 earlier are now incorporated.

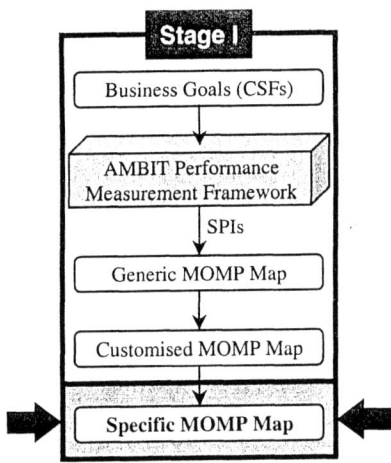

Figure 4.31 Develop the specific MOMP map

The instantiation of the customised MOMP map completes the transition from a *generic model* (Figure 4.4) through the *partial models* (generic MOMP maps for individual business processes in certain styles of manufacturing system), to the *particular model*, which represents the reality and unique characteristics of the particular manufacturing organisation. This particular MOMP map is prepared by customising the partial model (Section 4.6 above) to reflect particular sub processes peculiar to the organisation under analysis and finally adding organisation specific data as described above. An example of the particularisation of a MOMP is presented in the following Figure/Table.

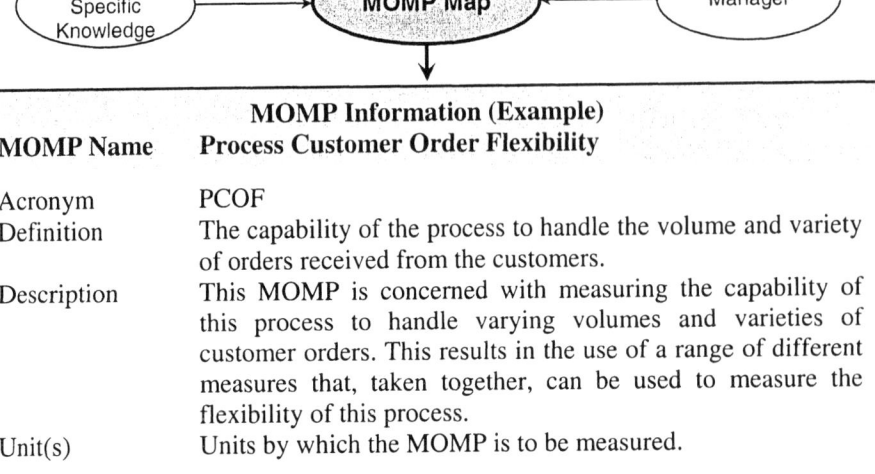

MOMP Information (Example)	
MOMP Name	**Process Customer Order Flexibility**
Acronym	PCOF
Definition	The capability of the process to handle the volume and variety of orders received from the customers.
Description	This MOMP is concerned with measuring the capability of this process to handle varying volumes and varieties of customer orders. This results in the use of a range of different measures that, taken together, can be used to measure the flexibility of this process.
Unit(s)	Units by which the MOMP is to be measured.

	MOMP Information (Example)
MOMP Name	**Process Customer Order Flexibility**
Measure(s)	▪ Number of orders received per unit time.
	▪ Number of orders processed per unit time.
	▪ Number of WIP orders per unit time.
	▪ Process throughput per unit time.
	▪ Percentage increase/decrease (+/-) in the order backlog.
	▪ Maximum, minimum and average number of orders received per unit time.
Target Value	Increase number of orders processed per unit time by 25%
Target Due Date	1 July 2004

4.8 Conclusion

In this chapter, a generic business model which represents at a high level, all of the activities occurring within a manufacturing organisation was defined. This AMBIT manufacturing model is based on the extended enterprise model of manufacturing systems.

A set of strategic performance indicators, which represent the dimensions along which an organisation can choose to compete in today's marketplace was also presented. These SPIs are based on:

▪ The generic manufacturing model and its associated business processes – customer order fulfilment, vendor supply chain, manufacturing, design co-ordination and co-engineering.

▪ The set of competitive performance measures – cost, quality, flexibility, time and the environment.

▪ The requirements of different types of manufacturing systems, namely make-to-stock, assemble-to-order, make-to-order and engineer-to-order systems.

In synopsis, Stage I of the AMBIT approach is concerned with the translation of the business goals, represented as critical success factors (CSFs) into a specific MOMP (measures of manufacturing performance) map directly relevant to the manufacturing organisation under analysis. The approach is based on the notion that ultimately a set of partial MOMP maps would be created and made available to the manager or management team and would be particularised to meet their specific requirements. The analysts would then select the appropriate MOMP maps based on the industry sector in which their manufacturing organisation is operating and the particular business process they wished to analyse. Following this, he would customise the business processes if necessary and complete the procedure by adding organisation specific data.

Chapter 5

The AMBIT Approach – Stage II

Keywords: Strategic management, knowledge about the technology and the programme, mechanics of the models, KTVs and KPVs, the MOMP-SPI-KTV/KPV linkage, modular structure of the models, core modules, relationships, technology component types, CE principles, CE characteristics, CE versus SE.

Chapter Objectives: The purpose of this chapter is to show the reader how we can link technology and programme issues to the manufacturing strategy. In doing so, we are able to assess the strategic implications of changes. Having read this chapter, the reader should become familiar with:

- The gap between strategic management and technology issues.
- The derivation of key technology (KTVs) and key programme variables (KPVs).
- Why strategic performance indicators (SPIs) are decomposed to measures of manufacturing performance (MOMPs), which are then further decomposed to key technology and key programme variables (KTVs/KPVs).
- The structure and mechanics of the technology and programme models.
- Sample KTVs and KPVs.

5.1 Introduction

As has been indicated earlier strategic decisions are typically long-term in nature and often fraught with unstructured and conflicting information. They are often based more on experience and intuition than on any analytical approach. Mintzberg [1994] suggests that strategic decision-making is more reliant on the thinking processes of the right cerebral hemisphere (by nature intuitive, holistic and spatial) than that of the left cerebral hemisphere (by nature analytical, rational and logical).

In contrast, manufacturing decisions are made by engineers, who generally approach problems from a technical and scientific background. As they tend to deal with specific data of a very technical nature, manufacturing decisions are more often than not, structured in nature. Examples of such decisions are equipment capability assessment, defect reduction and the implementation of a Design for Manufacturing (DFM) programme. It is this dichotomy in the nature of strategic decisions and manufacturing decisions (i.e. non-technical versus technical) which makes effective strategic manufacturing management difficult to achieve. The objective of Stage II therefore, is to bridge this knowledge gap (Figure 5.1) between strategic management and manufacturing technology.

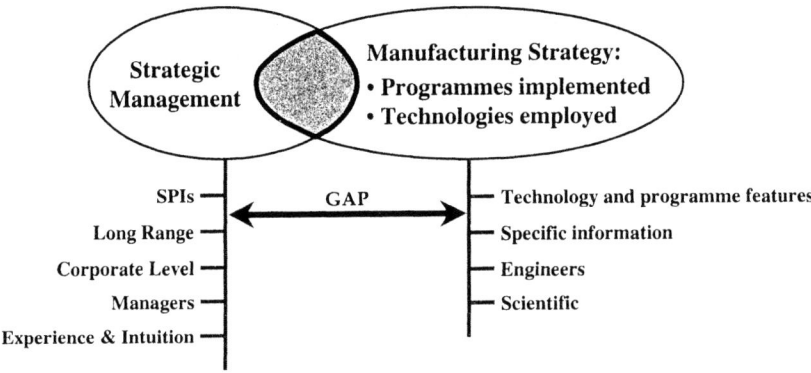

> Given the overriding significance of strategic decisions, it is imperative that they invariably result in successful outcomes.   – Harrison *et al.* 2001

Figure 5.1 The need to cater to the strategic management of technologies and programmes

Strategic decisions in manufacturing typically involve the selection of appropriate technologies and/or programmes and/or the implementation of particular technologies and/or programmes. We will refer to such decisions as technology decisions and programme decisions respectively.

1. A technology decision might involve the selection of the technologies to be employed to produce a particular end product or family of end products. For example:

    ▪ Casting of precision components, will involve decisions concerning some of the following: the selection of the manufacturing processes to be used, taking into account the advantages, disadvantages and costs associated with each process and the need to meet design and quality requirements in areas such as strength, tolerances, surface finish, etc.

    ▪ The selection of an appropriate interconnect technology to be used in electronics assembly. Interconnect technology refers to the various technologies used in the assembly of electronic components. The selection of such a technology impacts manufacturing in the widest sense, including the design function.

- Decision regarding the purchase of new computer numerically controlled machine tools and their subsequent integration into an existing manufacturing facility.

2. A programme decision typically relates to the organisation and to the logistics involved in the design, development and production of products. Examples include the introduction of World Class Manufacturing (see Chapter 3), the selection and implementation of an ERP (enterprise resource planning) or a PDM (product data management) system, the introduction of a new performance measurement system, the introduction of Lean Production or Concurrent Engineering.

Stage II of the AMBIT approach is concerned with the capture of knowledge about the technology and programme and its subsequent representation in a form suitable for decision making.

To illustrate the genericity of the AMBIT concept and framework, we will develop technology and programme models for interconnect technology and concurrent engineering respectively. There is no particular significance behind these choices; other technology and programme domains could have been used. The decisions that need to be made in the selection of the appropriate interconnect technology are complex and we believe that through the interconnect technology example we can elucidate the genericity of the AMBIT approach. Later in this chapter, we will present the basic terminology used in interconnect technology for readers not familiar with it.

In Stage II, a framework is developed which allows consideration of knowledge concerning indicators of the strategic performance of the organisation (i.e. SPIs), and knowledge relating to the key features of interconnect technology (i.e. key technology variables (KTVs)) and concurrent engineering (i.e. key programme variables (KPVs)). This framework is useful because it:

- Supports the decision-maker(s) in gaining a clearer understanding and appreciation of the technical issues of concern to the manufacturing function.
- Allows the decision-maker(s) to understand and articulate the impacts of the introduction of and/or changes to a technology or programme.
- Provides the necessary structure so that analysis tools (Stage III of AMBIT approach – Chapter 6) can be used to evaluate the consequences of changes proposed.

Kakabadse *et al.* suggest that to successfully envision the future, all the members of the top team need to be involved. "The top team as a whole must share the same vision for the future, while still making individual inputs and variations, so that strategies, structures and cultures can be created and maintained to achieve that vision" [Kakabadse *et al.* 1995]. We believe that the AMBIT approach supports a shared vision, through its bridging of the gap between management and engineers and the technical personnel.

The specific aim of Stage II is to study the MOMP map generated in Stage 1 and analyse the suggested technologies or programmes in terms of their direct impact on these MOMPs. Elements of the MOMP map define precisely what measures of manufacturing performance need to be addressed in order to achieve the strategic goals of the organisation. Key technology variables (KTVs) and key programme variables (KPVs) are used to differentiate the technologies and programmes respectively. The KTVs and KPVs represent key features of the technology or programme under consideration. In other words, they are used to differentiate the various options available and more critically, they are capable of being expressed in terms of their projected impact on individual MOMPs.

The outputs of Stage II are KTV/KPV templates that essentially describe the technologies or programmes under consideration in terms of their impact on the manufacturing organisation as captured and articulated by the measures of manufacturing performance. These templates result when the knowledge concerning the selected technology and programme (expressed through KTVs and KPVs respectively) is mapped to the measures of manufacturing performance (MOMPs) on the bottom level of the specific MOMP map. Since the connectivity between Stages I and II is through the MOMPs, we can state that the specific MOMP set acts as a 'bridge' between the business strategy and the technology and programme models (Figure 5.2).

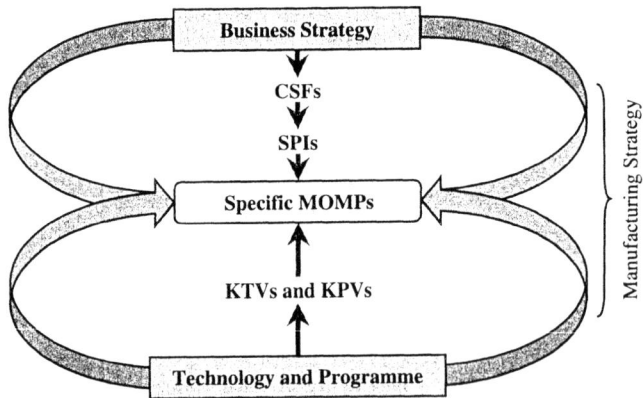

Figure 5.2 The linkage of the business strategy to the technology and programme models

Stage II of the AMBIT approach therefore comprises the following steps (Figure 5.3):

- **The documentation of knowledge about the technology and programme**: The AMBIT approach seeks to support managers in analysing how changes to an existing technology or programme or the implementation of a new technology or programme will impact on the organisation and the achievement of its overall strategic aims. However, for the analysis to be as effective as possible, it is necessary to have up-to-date, accurate information, so that realistic models of the technology and programme domains can be built. To achieve this, knowledge about the relevant technology and programme must be captured.

- **The creation of a generic technology/programme model**: After knowledge capture, models of the relationship between the technological and programme issues and the specific MOMPs (defined during Stage I) are built. Key technology variables (KTVs) represent key product and/or process features, which differentiate, for example, one interconnect technology from another. Key programme variables (KPVs) represent the essential features of a proposed programme: for example features that clearly differentiate concurrent and sequential engineering. These KTVs or KPVs are expressed in terms of their projected impact on manufacturing performance via the MOMPs.
- **The creation of a specific model (KTVs and KPVs)**: As with Stage I of the AMBIT framework, the generic technology/programme model is edited by the user, in order to modify or remove key technology variables (KTVs) and/or key programme variables (KPVs), which are regarded as inappropriate to the organisation in question and to capture enterprise specific data on those variables.

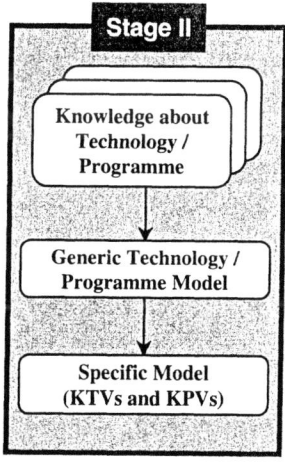

Figure 5.3 The AMBIT approach - Stage II

As an aside, it is worth noting that the technology and programme models developed in Stage II also support a high degree of learning by the decisions makers and management team, which is certainly supportive of good decision making.

5.2 Knowledge about the Programme and the Technology

There are three basic steps to strategic decision-making: getting the right information, making a good decision and then implementing that decision. Success in the information stage translates into knowing the types of information needed to make the decision, finding it, and transmitting it quickly to the decision-makers.   – McNeilly 2002

Today's electronics organisations are continuously striving for smaller, faster, cheaper and more reliable products with increased functionality. As a result, changes in the current interconnect technology used by an organisation or indeed investments in new interconnect technologies, can produce technological dislocations that can be threats to the organisation's survival or opportunities for its future growth. In a similar way, changes to existing manufacturing programmes or the introduction of new manufacturing programmes can have repercussions that can either impede or facilitate success. Thus, prior to any decision concerning existent technology and programme modification and/or new technology and programme introduction, it is first necessary to build realistic models of the technology and programme domains. Whilst these models are necessary from the point of view of analysis, they also help management learn more about the domain in which the decisions are to be made.

Before such models (Figure 5.4) can be built, knowledge concerning the relevant technology and programme domains needs to be captured. This knowledge capture process is usually carried out through extensive and intensive literature surveys and interviewing industrial practitioners directly involved in the key technology and programme domains of interest - in our case, Interconnect Technology and Concurrent Engineering.

The knowledge and experience gained from literature survey (academic and technical press and product data sheets and documentation) and the interviews (i.e. discussions with "best practice" users of the technology) helps in:
- Providing a description of all elements of each technology.
- Identifying Key Technology Variables (KTVs) and Key Programme Variables (KPVs).
- Forming logical paths (linkages) from KTVs to MOMPs and from KPVs to MOMPs.

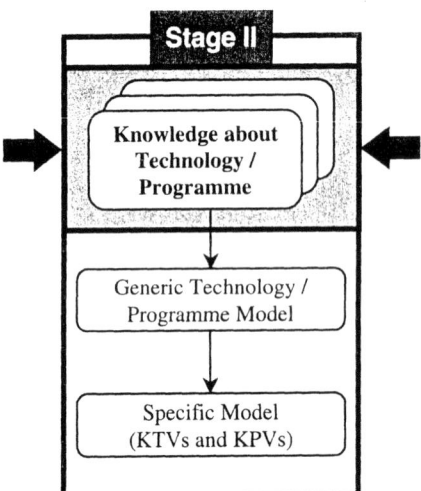

Figure 5.4 Progress in Stage II – knowledge about technology and programme

This knowledge can then be used to develop models of the impacts of the technological and programme issues on manufacturing performance. If inaccurate information is obtained concerning the domain under study, then it goes without saying that the final analysis will not reflect the true situation. The value and effectiveness of the analysis is dependent on access to timely and correct information.

5.3   Generic Programme and Technology Models

Once the knowledge concerning interconnect technology and concurrent engineering is captured, semantic models of the relationships between the technological and programme issues and the upper levels of the manufacturing strategy can be developed. These semantic models (Figure 5.5) describe the strategic impact of changes in key technology variables and key programme variables for the interconnect technology and the concurrent engineering domains respectively. As we have already stated, key technology variables (KTVs) represent key product and/or process features, which differentiate one interconnect technology class from another. Key programme variables (KPVs) represent the essential features of the product development process, for example features whose parameters differentiate Concurrent and Sequential engineering.

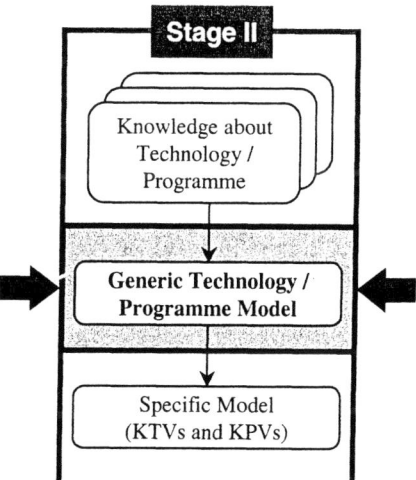

Figure 5.5 Progress in Stage II – generic technology and programme model

5.3.1   The Background to the Models

As has been indicated, the KTVs and KPVs are essentially characteristics which distinguish or differentiate a technology or programme in terms of its influence on a measure of manufacturing performance (MOMP). We will now explore this linkage of the KTVs / KPVs with MOMPs in more detail.

Measures of Manufacturing Performance (MOMPs)

In Stage I of the AMBIT approach, we saw that the goals of the organisation are initially expressed in terms of its critical success factors (CSFs). The AMBIT performance measurement framework is then used to derive a set of business metrics, which measure the strategic performance and progress of that organisation, with respect to the satisfaction of its business goals. These metrics are termed strategic performance indicators (SPIs). As outlined in chapter 4, the SPIs are decomposed into generic MOMP maps. This map is customised and instantiated in order to reflect the individual characteristics of the organisation in question. Customisation and instantiation of the generic MOMP maps result in the specific MOMP maps.

In order to measure how changes in the technology and programme impact the business strategy, it is necessary to link the model of Interconnect Technology and Concurrent Engineering to the business strategy via the specific MOMP map developed in Stage I. This is achieved by articulating the impact of the Key Technology or Programme Variables on the MOMPS (see Figure 5.6).

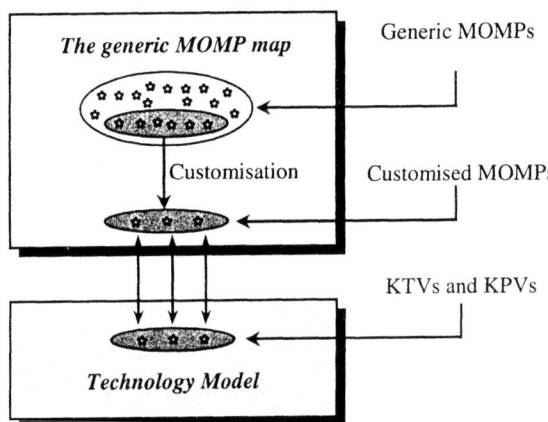

Figure 5.6 The link between the specific MOMPs and the technology model

As shown in chapter 4, the specific MOMPS of Stage I are categorised according to the business processes in the AMBIT performance measurement framework (i.e. customer order fulfilment, manufacturing, design co-ordination, co-engineering, and vendor supply) and the macro measures of competitive performance (i.e. cost, quality, flexibility, time and environment) within these processes. Clearly the KTVs and KPVs need to be examined in terms of their possible impact on particular business processes in order to understand which MOMP maps they impact.

In the case of Interconnect Technology, the introduction of a new printed circuit board (PCB) assembly technology affects some of the processes more than others. For example, the manufacturing process is highly influenced by changes in the electronic assembly technology, whilst the co-engineering process is affected in a minor way. Because many of the co-engineering activities are independent of the assembly technology, the co-engineering process is omitted from the technology model. Meanwhile, as concurrent engineering (CE) is primarily concerned with the product development process; the main business processes affected by the introduction of a CE programme are the design co-ordination and the co-engineering processes respectively.

Key Technology Variables (KTVs) and Key Programme Variables (KPVs)

As has been indicated, a key technology variable (KTV), such as the placement rate in the case of electronics assembly technology, basically describes a critical or *key* product or process feature, which allows the user to differentiate one technology class from another. For example a KTV can be used to differentiate between surface mount technology (SMT) and chip on board (COB) (in the case of interconnect technology). A key programme variable (KPV) represents a critical feature which differentiates one programme from another. In the case of concurrent engineering and sequential engineering, examples of KPVs include:
- The organisation of work.
- The problem solving approach.
- The style of project leader.

Table 5.1 shows some of the differentiating characteristics of KTVs and KPVs.

Table 5.1 Differentiating characteristics of KTVs and KPVs[10]

	KTVs	**KPVs**
Type	Quantitative	Qualitative
Objectivity	Objective	Somewhat more subjective
Associated Processes	Manufacturing, customer order fulfilment, vendor supply and design co-ordination	Co-engineering and design co-ordination
Technologies / Programmes	Plated though hole (PTH), surface mount technology – single chip (SMT-SC), fine pitch technology (FPT), chip on board (COB) and multi chip modules (MCM)	Concurrent engineering and sequential engineering
Focus	Capital equipment investment	Investment in human endeavour, new work processes.

A key point to highlight is the fact that the introduction of a new or changes to an existing technology and/or programme results in changes to the corresponding KTVs and/or KPVs. Furthermore, these changes are reflected in changes in appropriate MOMPs (Figure 5.7). Through the MOMPs, the effects of changes in

---

[10] Using the examples of Interconnect Technology and Sequential / Concurrent Engineering.

the KTVs and KPVs on the manufacturing system, map back to the SPIs and ultimately to the Critical Success Factors. Thus, the impact on business strategy can be articulated and partially if not completely quantified (see Figure 5.7).

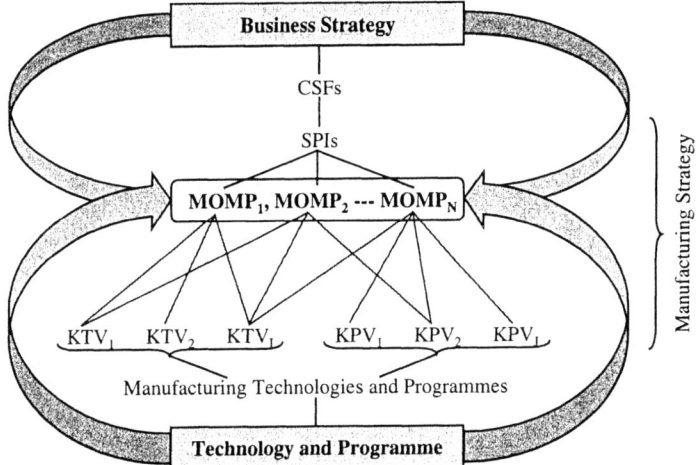

Figure 5.7 Changes in KTVs and/or KPVs influence the MOMPs

The high level nature of a SPI makes the identification of its value in relation to a technology and/or programme class difficult; hence the need to consider the impact of KTVs and KPVs on MOMPS as illustrated in Figure 5.8.

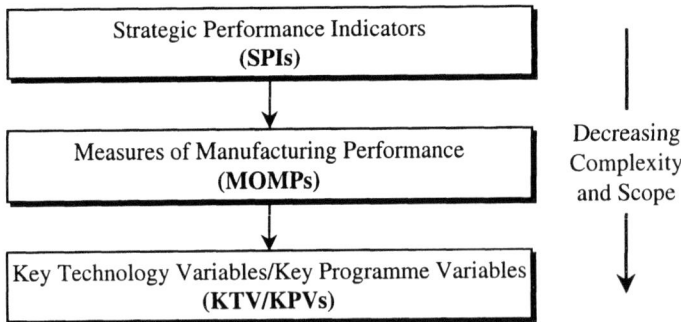

Figure 5.8 The SPIs→MOMPs→KTV/KPVs decomposition process

A simple example that highlights the difficulty of SPI quantification is shown in Figure 5.9. The determination of the value for a particular assembly technology of the 'Time Consumed by the Manufacturing Process' SPI is difficult without detailed knowledge of the physical implementation of that actual technology.

The Time Consumed by the Manufacturing Process SPI has to be decomposed into a set of supporting MOMPs (e.g. queue time of each sub-assembly process, processing time of sub-assembly process and transport time of sub-assembly process). The values of these MOMPs are obtained from the KTVs of the technologies; for example in the case of the processing time for the insertion or 'onsertion' sub-assembly process the insertion or placement rate is the appropriate KTV.

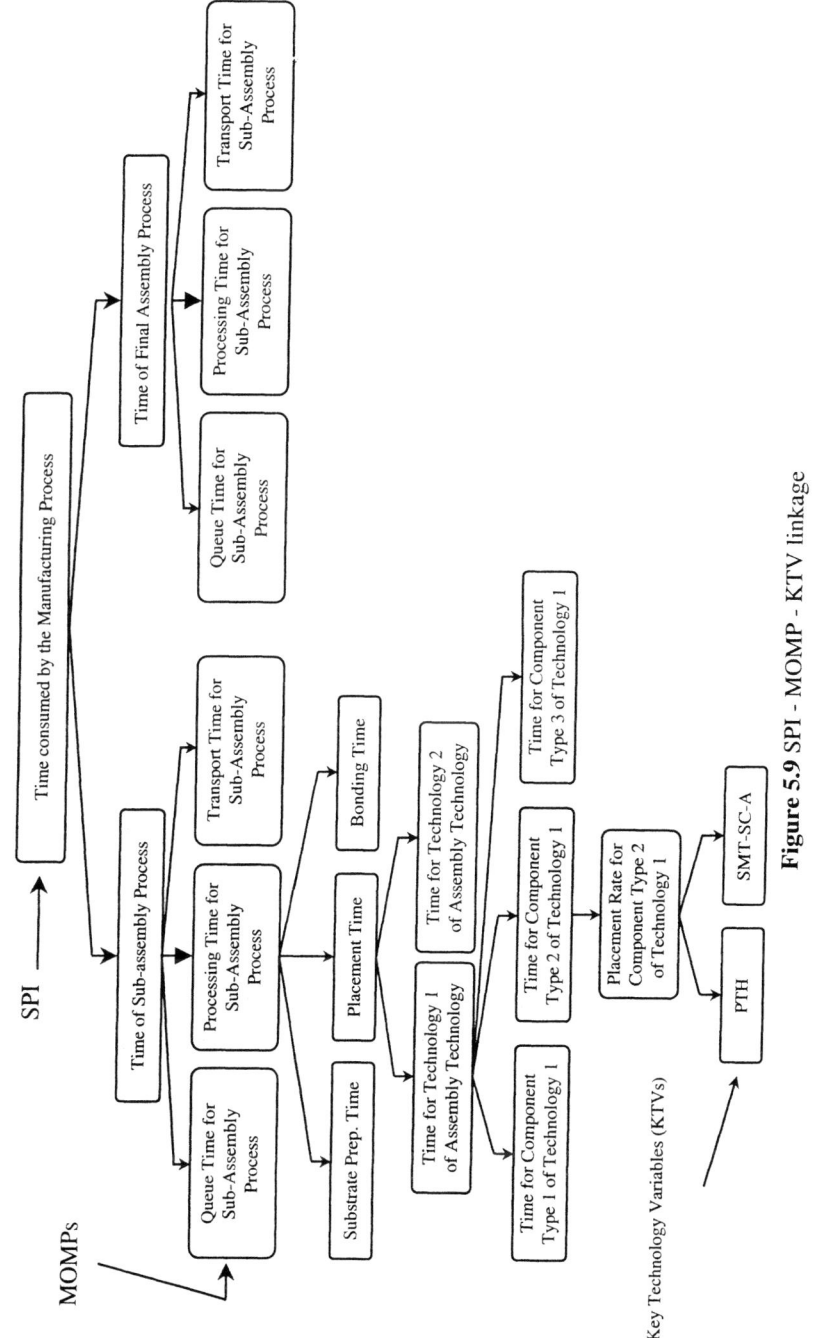

Figure 5.9 SPI - MOMP - KTV linkage

5.3.2 The Structure of the Models

The technology model shows the relationship (or causal linkages) between the key technology variables (KTVs) and the technology related measures of manufacturing performance (MOMPs). Meanwhile, the programme model shows the relationship (or causal linkages) between the key programme variables (KPVs) and the MOMPs. Both models are composed of objects and relationships between objects. By browsing each model, a manager can gain a better understanding of how current and future technologies and programmes impact the organisation. Figure 5.10 highlights the generic and modular nature of the technology and programme models. Figures 5.11 and 5.12 outline the structure of the technology model and the programme model respectively.

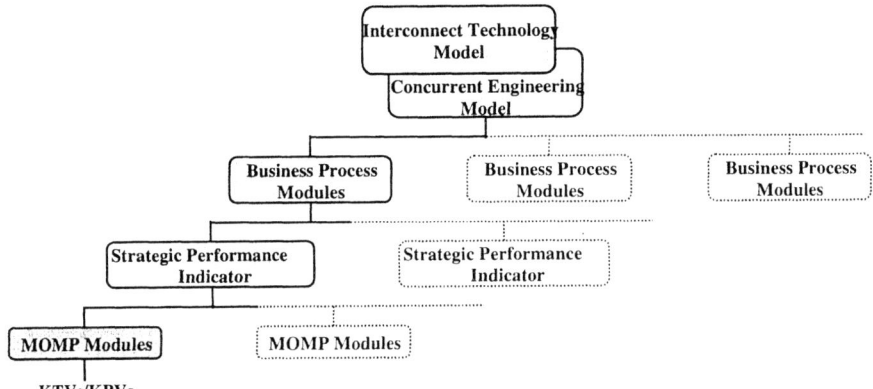

Figure 5.10 The modular nature of the models' structure

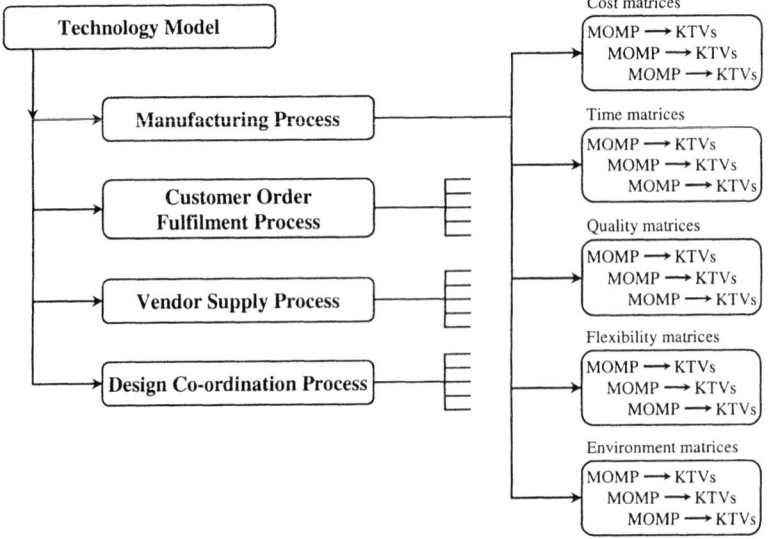

Figure 5.11 Structure of the technology model.

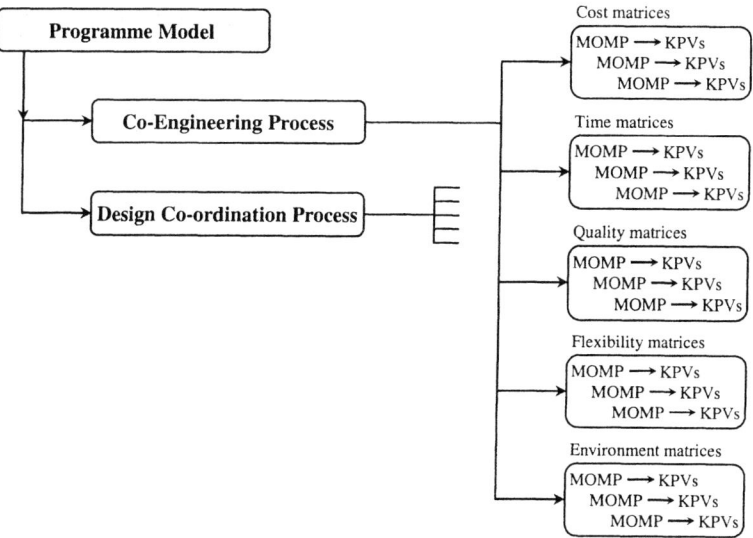

Figure 5.12 Structure of the programme model

It is evident from this figure that both models are composed of the business process modules, the strategic performance indicators and the MOMP modules. In the interconnect technology model, the business process modules include the manufacturing process, vendor supply process, design co-ordination process and the customer order fulfilment process modules (the co-engineering process has been omitted due to its minimal influence to the technology decision to be made). In the concurrent engineering programme model, the business process modules include the design co-ordination process and the co-engineering process modules.

Each of the process modules contains a module for the specific SPI (cost, time, quality, flexibility and the environment) affected by the technology or programme. These in turn are decomposed into MOMP maps of the particular MOMPs affected by the technology or programme.

We will now go on to apply these ideas to the development of a technology model for Electronics Assembly, using advanced Interconnect Technology.

5.4   A Generic Technology Model for Interconnect Technology in Electronics Assembly

As indicated, the technology model is concerned with MOMPs and KTVs and specifically the impact of the KTVs on the MOMPs.

The Measures of Manufacturing Performance.

In Table 5.2, examples of some of the high level MOMPs used in the electronics industry and categorised under the macro measures of performance for the relevant business processes are presented.

Table 5.2 Examples of MOMPs for the technology model in electronics assembly

Process: Manufacturing

Cost	▪ Cost per insertion
	▪ Conversion burden
Quality	▪ Yield at key points
	▪ True yield
Flexibility	▪ Process cycle time per product
	▪ Average set-up time per unit
Time	▪ Process cycle time
	▪ Inventory turns
Environmental Impact	▪ Level of toxic waste per product
	▪ Level of toxic waste per process

Process: Vendor Supply

Cost	▪ Cost of inventory
	▪ Cost of out of stock
Quality	▪ Number of on-time deliveries
Flexibility	▪ Average delivery lead time
	▪ Total number of vendors
Time	▪ Average delivery lead time
	▪ Number of on-time deliveries
Environmental Impact	N/A

Process: Design Co-ordination

Cost	▪ Cost per design version
	▪ Average development cost per product
Quality	▪ Number of design changes
	▪ 1st pass yield
Flexibility	▪ Average hours per drawing
	▪ Level of modularisation
Time	▪ Average hours per drawing
	▪ Time to market for product
Environmental Impact	▪ Percent of material re-cyclable
	▪ Level of toxic waste per product

Process: Customer Order Fulfilment

Cost	▪ Customer order manufacture cost
Quality	▪ Customer order manufacturing quality
Flexibility	▪ Process cycle time
	▪ Average set-up time of process
Time	▪ Delivery time
	▪ Inventory Turns
Environmental Impact	▪ Percent of material re-cyclable
	▪ Level of toxic waste per product

As already outlined, each MOMP is decomposed in order to be able to capture the effects of the KTVs, which influence that MOMP. Before doing this we must categorise each KTV according to the relevant business processes and the macro measures of competitive performance within those processes which it impacts (see Figure 5.13).

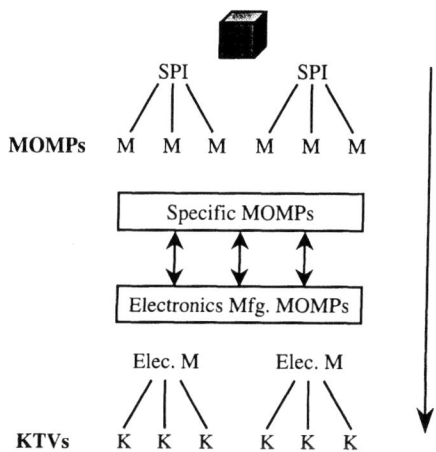

Figure 5.13 Identification of the KTVs with the MOMPs

The Key Technology Variables (KTVs)

In Tables 5.3, 5.4 and 5.5, we indicate the KTVs which impact the various business processes, influenced by interconnect technology. These business processes are; the manufacturing process, the vendor supply process and the design co-ordination process. The KTVs for each of these processes are categorised by the macro measures of competitive performance (cost, time and quality).

The information contained in these tables is not exhaustive and is presented for illustration purposes only. For example, two macro measures of competitive performance, i.e. flexibility and the environment have not been included. If, in a real-life scenario these measures were considered to be critical, they would be included and their KTV categorisation would follow an approach similar to that of cost time and quality. Similarly, two manufacturing business processes (customer order fulfilment and co-engineering processes), which are omitted in Tables 5.3 through 5.5, would also be categorised, if they were considered important.

Table 5.3 KTVs in the manufacturing process (in electronics assembly) categorised by macro measures of competitive performance

---

Manufacturing Process – Macro Measure: Cost

1.   Substrate preparation cost
      1.1   Substrate preparation equipment cost
      1.2   # People assigned to substrate preparation
      1.3   Cost per unit of substrate preparation material
      1.4   Substrate fixture development cost
      1.5   Average substrate preparation process time (influences cost of substrate preparation process)
      1.6   Average substrate preparation set-up time (influences cost of substrate preparation process)
2.   Component placement cost
      2.1   Component placement equipment cost
      2.2   # People assigned to component placement
      2.3   Average component placement set-up time (influences cost of component placement process)
      2.4   Average component placement rate (influences cost of component placement process)
3.   Bonding cost
      3.1   Bonding equipment cost
      3.2   # People assigned to bonding process
      3.3   Cost per unit of bonding material
      3.4   Average bonding process time (influences cost of bonding process)
      3.5   Average bonding set-up time (influences cost of bonding process)
4.   Cleaning cost
      4.1   Cleaning equipment cost
      4.2   # People assigned to cleaning process
      4.3   Cost per unit of cleaning material
      4.4   Average cleaning time (influences the cost of cleaning process)
      4.5   Average cleaning set-up time (influences the cost of cleaning process)
5.   Testing cost
      5.1   Testing equipment cost
      5.2   Testing fixture development cost
      5.3   # People assigned to test
      5.4   Average in-circuit testing time (influences cost of testing)
      5.5   Average in-circuit set-up time (influences cost of testing)
6.   Rework and repair cost
      6.1   Rework & repair equipment cost
      6.2   # People assigned to rework and repair
      6.3   Time to rework a component (influences cost of rework process)
      6.4   Time to repair a component (influences cost of repair process)

Manufacturing Process – Macro Measure: Time

1.   Substrate preparation time
      1.1   Average substrate preparation process time
      1.2   Average substrate preparation set-up time
2.   Component placement time
      2.1   Average component placement rate
      2.2   Average component placement set-up time
3.   Bonding time

       3.1   Average bonding process time
       3.2   Average bonding set-up time
4.   Cleaning Time
       4.1   Average cleaning time
       4.2   Average cleaning set-up time
5.   Test time
       5.1   Average in-circuit testing time
       5.2   Average in-circuit set-up time
       5.3   Time to repair a component
6.   Rework and Repair time
       6.1   Time to rework a component
       6.2   Time to repair a component

Manufacturing Process – Macro Measure: Quality

1.   Quality of substrate preparation process
       1.1   Machine repeatability in substrate preparation
       1.2   Degree of material deposition accuracy in substrate preparation
       1.3   Degree of material volume accuracy of substrate preparation machine
       1.4   # of variables in substrate preparation accuracy
       1.5   Component lead pitch
       1.6   Number of component leads
       1.7   Component has a package - Y/N?
       1.8   Component size
2.   Quality of component placement process
       2.1   Degree of placement machine repeatability
       2.2   Degree of component placement position accuracy
       2.3   Degree of component placement pressure accuracy
       2.4   Component lead pitch
       2.5   Number of component leads
       2.6   Component has a package - Y/N?
       2.7   Component size
       2.8   Component reliability
3.   Quality of bonding process
       3.1   Degree of temperature control during bonding
       3.2   Degree of accuracy of bond
       3.3   Component lead pitch
       3.4   Number of component leads
       3.5   Component has a package - Y/N?
       3.6   Component size
       3.7   Component reliability

Table 5.4 KTVs in the vendor supply process (in electronics assembly) categorised by macro measures of competitive performance

---

Vendor Supply Process – Macro Measure: Cost

1.   Material Availability
       1.1   Component availability
       1.2   Substrate availability
       1.3   Supplier lead time
2.   Cost of Material

2.1   Average cost of component
2.2   Average cost of substrate
2.3   Cost per unit of substrate preparation material
2.4   Cost per unit of bonding material
2.5   Cost per unit of cleaning material

Vendor Supply Process – Macro Measure: Time

1.   Material Availability
1.1   Component availability
1.2   Substrate availability
1.3   Supplier lead time

Vendor Supply Process – Macro Measure: Quality

1.   Component technical features
1.1   Component reliability

Table 5.5 KTVs in the design co-ordination process (in electronics assembly) categorised by macro measure of competitive performance

---

Design Co-ordination Process – Macro Measure: Cost

1.   Availability of tools
1.1   CAD tool availability
1.2   Availability of design for assembly tools
1.3   Availability of design for testability tools
1.4   Availability of thermal management design tools

Design Co-ordination Process – Macro Measure: Time

1.   Availability of tools
1.1   CAD tool availability
1.2   Availability of design for assembly tools
1.3   Availability of design for testability tools
1.4   Availability of thermal management design tools

Design Co-ordination Process – Macro Measure: Quality

1.   Component Technical Features
1.1   Component lead pitch
1.2   Number of component leads
1.3   Component size
1.4   Component has a package - Y/N?
1.5   Component height profile
1.6   Component reliability
2.   Substrate Technical Features
2.1   Substrate line width
2.2   Substrate wiring density

A sample list of KTVs in the electronics assembly is presented in Table 5.6.

Table 5.6 A sample list of KTVs (*in two columns*) in Electronics Assembly

- Substrate fixture development cost
- Average cost of component
- Cost per unit of cleaning material
- Average cleaning time
- Average cleaning set-up time
- Testing equipment cost
- Testing fixture development cost
- Average component placement rate
- Substrate preparation equipment cost
- Bonding equipment cost
- Average bonding process time
- Cleaning equipment cost
- # People assigned to test
- Average in-circuit testing time
- Average in-circuit set-up time
- Time to rework a component
- Time to repair a component
- Cost per component
- Cost per substrate
- Rework & repair equipment cost
- # skilled people assigned to substrate preparation
- Average substrate preparation set-up time
- Availability of design for assembly tools
- # skilled people assigned to component placement
- Average component placement set-up time
- # skilled people assigned to bonding process
- Degree of material volume accuracy of substrate preparation machine
- # of variables in substrate preparation accuracy
- Degree of component placement position accuracy
- # skilled people assigned to cleaning process
- # People assigned to rework and repair

- Supplier lead time
- Average cost of substrate
- Cost per unit of bonding material
- Degree of accuracy of bond
- Number of component leads
- Component reliability
- Component lead pitch
- CAD tool availability
- Component reliability
- Cost per unit of bonding material
- Average bonding set-up time
- Cost per unit of cleaning material
- Number of component leads
- Component size
- Component has a package - Y/N?
- Substrate line width
- Substrate wiring density
- Component availability
- Substrate availability
- Component height profile
- Cost per unit of substrate preparation material
- Component placement equipment cost
- Availability of design for testability tools
- Availability of thermal management design tools
- Average substrate preparation process time
- Machine repeatability in substrate preparation
- Degree of material deposition accuracy of substrate preparation
- Degree of placement machine repeatability
- Degree of temperature control during bonding
- Degree of component placement pressure accuracy
- Cost per unit of substrate preparation material

5.4.1 Linking the KTVs to the MOMPs

In order to develop this part of the model we need to consider some of the technical details of the interconnect technologies used in electronics assembly. This we will do briefly now.

An Overview of Interconnect Technology

The mechanics of assembling electronic components to a base material is termed interconnect technology. In order to familiarise the reader with the terms employed while developing the technology model in this chapter, it is useful to outline these terms here. The continuum of electronics assembly contains five primary assembly technologies, as described in Table 5.7.

Table 5.7 Primary interconnect technologies

1.	PTH	Plated through hole.
2.	SMT-SC	Surface mount technology – single chip.
3.	FPT	Fine pitch technology.
4.	COB	Chip on board.
5.	MCM	Multi chip modules.

The phrase "assembly technology" is used to describe the assembly of a printed circuit board (PCB) using either a single primary technology or a mix of two primary technologies (e.g. the assembly technology may be a combination of PTH and SMT-SC). Examples of possible assembly technologies are shown in Table 5.8

Table 5.8 Some examples of assembly technologies

Assembly Technology ↓	Technology	
	1	2
Plated through hole (PTH): Printed circuit board (PCB) with only plated through hole (PTH) components on one side of the substrate. The other side does not hold any components.	PTH	–
SMT-SC-A: PCB with only surface mount technology (SMT) components on one side of the substrate. The other side does not hold any components.	SMT-SC	–
SMT-SC-B: PCB with only surface mount technology – single chip (SMT-SC) components on both sides of the substrate.	SMT-SC	–
SMT-SC-C: PCB with SMT-SC and PTH components on one side of the substrate, and SMT-SC components on the other side.	PTH	SMT-SC
Chip on board - A (COB-A): PCB with chips on one side of the substrate. The other side does not hold any components.	COB	–
Chip on board - B (COB-B): PCB with chips and SMT-SC components on one side of the substrate, and chips on the other side.	COB	SMT-SC

If a manufacturer's PCB assembly technology is a mix of two primary technologies, then the two technologies which form the mix are referred to in the model as *Technology 1* and *Technology 2* respectively. For example, if the assembly technology uses a combination of plated through hole (PTH) and surface mount technology – single chip (SMT-SC), then Technology 1 refers to PTH, whilst Technology 2 refers to SMT-SC. In order to maintain consistency in the technology model, a single technology is treated as a mix of technologies, with Technology 1 referring to the single technology and Technology 2 acting as a redundant term (i.e. it does not refer to any technology). As both sides of the substrate can be used in surface mount technology – single chip (SMT-SC), fine pitch technology (FPT), chip on board (COB), and multi chip modules (MCM), the number of assembly technology options within the model increases.

In some assembly technologies involving a mix of primary technologies, Technology 1 and/or Technology 2 may be implemented differently on each side of a substrate. An example of this is in SMT-SC-C. Before the placement of SMT components on side 1, adhesive dots are deposited at different component locations on the substrate. On side 2, the substrate is prepared by screen-printing. In this particular case, Technology 1 on side 1 is regarded differently to Technology 1 on side 2. Technology 1 and Technology 2 are normally viewed, where appropriate, in terms of *the component types* of the primary technologies they represent (see Table 5.9). For example, the component types for plated through hole (PTH) are axial, radial and dual inline packages (DIP). Surface mount technology - single chip (SMT-SC) and fine pitch technology (FPT) have two component types: passive chips and active components, whilst chip on board (COB) and multi chip modules (MCM) have only one component type, the bare Integrated Circuit (IC) chip. As already stated, the terms Technology 1 and Technology 2 refer to the primary technologies that comprise an assembly technology. The component types of these technologies are referred to as Component Type 1, 2 or 3.

Table 5.9 The technology component types

Assembly Technology ➜	PTH	SMT-SC-A	SMT-SC-B	SMT-SC-C	COB-A	COB-B
Technology 1	PTH	SMT-SC	PTH	PTH	COB	COB
Component Type 1	Axial	Passive	Axial	Axial	Chip	Chip
Component Type 2	Radial	Active	Radial	Radial	-	-
Component Type 3	DIP	-	DIP	DIP	-	-
Technology 2	-	-	SMT-SC	SMT-SC	-	SMT-SC
Component Type 1	-	-	Passive	Passive	-	Passive
Component Type 2	-	-	Active	Active	-	Active
Component Type 3	-	-	-	-	-	-

Component Type 1 in PTH refers to axial components whilst *Component Types 2 and 3* refer to radial and dual inline package (DIP) type components respectively. In surface mount technology - single chip (SMT-SC), *Component Type 1* refers to passive components whilst *Component Type 2* refers to active components.

However, as there is no third component type in SMT-SC, Component Type 3 is redundant. Table 5.9 shows how the technology model represents each assembly technology and its component types.

Key technology variables (KTVs - Table 5.10 provides indicative values for those KTVs) that can influence the electronics assembly include:

- *Lead Pitch.* The component lead pitch is the spacing between neighbouring lead centres. Lead pitch is measured in mils (thousandths of an inch).
- *Component Density.* The component density refers to the number of components that can be placed per unit area of board. It is measured in terms of the number of VLSI chips per square inch.
- *Board Silicon Density.* This is the percentage of the board area occupied by silicon. It is expressed as a percentage, and is a measure of the depth of functionality of the assembled board.
- *Number of Component Leads.* This represents the number of leads on the component. It serves as a measure of the functionality of the component in the sense that more leads imply more communication with other devices.
- *Component Size.* The component size measures the level of component complexity and the logic density of the device. The smaller the size the more components can be placed in the same area, and the greater the importance of management of heat dissipation. The component size is measured in terms of that of a typical Plated Through Hole (PTH) Integrated Circuit.
- *Package Presence.* The package is the plastic housing that serves as a protector and interconnection media for the device. It has important implications for size, handling, environmental control and thermal management.
- *Substrate Wiring Density.* The substrate wiring density refers to the amount of interconnect wiring per unit area per layer of substrate media (usually a printed circuit board (PCB)). The greater the substrate wiring density, the more interconnection that can be placed in a smaller area and the smaller the system. This variable is measured in terms of inches of per square inch per layer.

Table 5.10 Sample values for the KTVs

Sample Values for the KTVs					
KTV	PTH	SMT	FPT	COB	MCM
Lead Pitch (mm)	100	50-31 50-25	25-15	15-4	8-4
Component Density (VLSI chips)	0.3	0.5 (approx.)	0.5-1	1-1.5	4
Board Silicon Density	4%-6%	5%-10%	~ 10%	~ 10%	50%-80%
Number of Component Leads	3-132	16-68-132	40-224	80-300	100-300-500
Component Size (relative to PTH)	1	0.8-0.5	0.5	0.25 (bare die)	0.25 (bare die)
Package Presence	yes	yes	yes	no	no
Board Wiring Density (in./sq. in./layer)	30-40	40-70	50-70	50-70	250-300

In the case of the interconnect technology model, the linkage between the MOMPs and KTVs is described by firstly representing the linkage in terms of two primary technologies: Technology 1 and Technology 2. Because we are looking here at a two sided printed circuit board (PCB) the linkage is also described by the way the primary technologies are implemented on each side of the substrate or board. Finally, the linkage is represented in terms of the way each component type, for each primary technology, is accommodated on each side of the substrate (Figure 5.14).

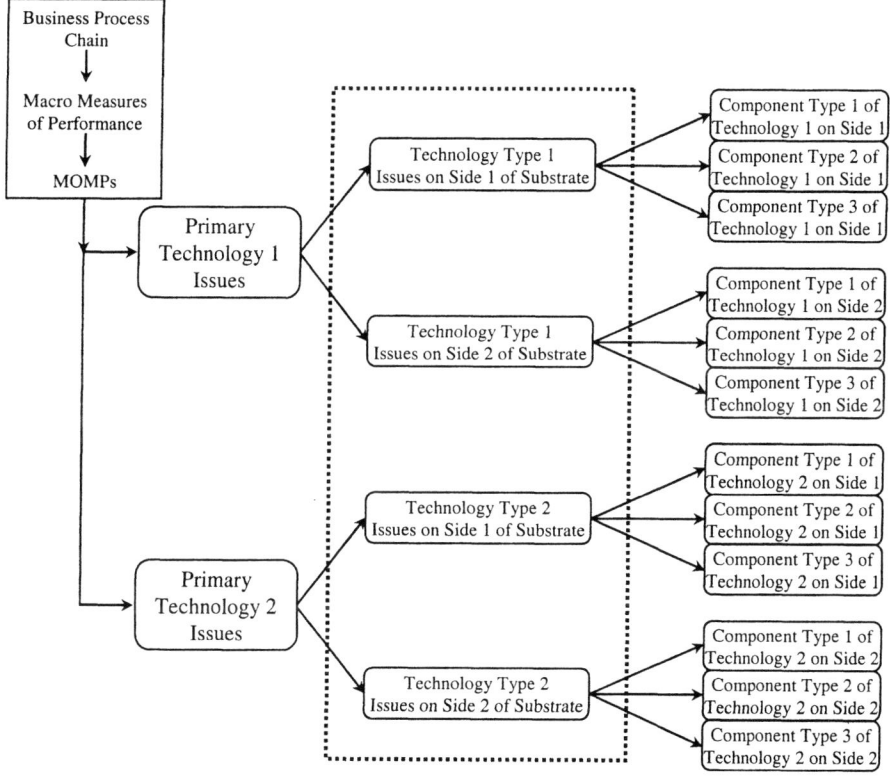

Figure 5.14 Structure of the linkage

The linkage structure (shown in Figure 5.14) is used to represent the modules in the manufacturing, customer order fulfilment and design co-ordination processes. However, the modules of the vendor supply process are represented differently, as each primary technology in the vendor supply process is not viewed in terms of the sides of the substrate on which it is implemented.

Knowledge representation in the Models

The knowledge within the technology and programme models is represented using a frame-based approach, which reflects to a large degree the way humans view and store knowledge about objects. In this approach, frames are used to organise knowledge in a way that directs attention and facilitates recall and inference. Frames describe information about an object and also about the object's relations. Furthermore, two or more objects can be related to each other to form a network of frames (see Figure 5.15). Whilst Figure 5.15 represents the linkage of the MOMP 'True Yield' to the KTVs (Quality of Type 1 components, and Performance of placement equipment for type 1 components of technology 1) which influence it, it also represents a network of frames. One such frame is the True Yield frame. Another frame is Number of PCBs without Rework. Each frame contains information relating to its corresponding object i.e. in this case, True yield and Number of PCBs without Rework.

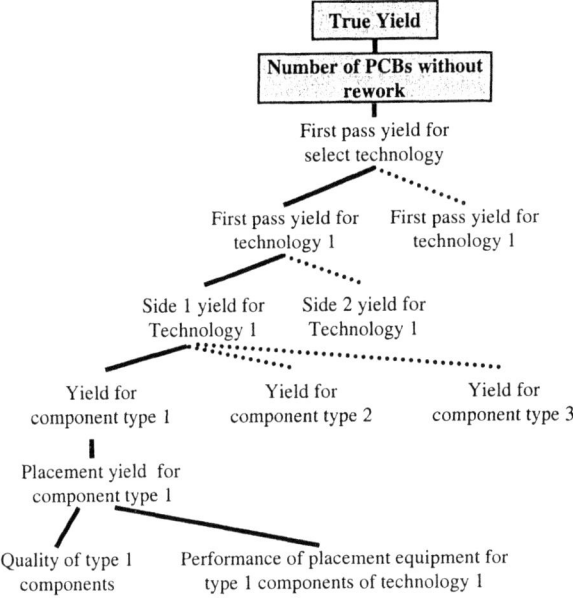

Figure 5.15 A network of frames

As already stated, key technology variables (KTVs) basically describe critical or *key* product or process features. These features allow the user to differentiate one interconnect technology class from another (i.e. they support the user in differentiating between Surface Mount Technology (SMT) and Chip On Board (COB) technologies). KTVs vary not only between different technologies but also between different printed circuit board styles within the same technology. Consider Figure 5.9 discussed earlier where KTVs are used to differentiate between printed circuit board (PCB) assembly technologies on three levels, as follows:

- **Level 1**: The KTVs refer to the printed circuit board (PCB) assembly technology in general, e.g. availability of CAD tools, cost of testing equipment etc.
- **Level 2**: The KTVs refer to the component categories of Technology 1 or Technology 2 within an assembly technology, independent of the side of the substrate the primary technology is on. Examples include:

o   Placement machine cost for component type-1 of Technology 1.
o   Placement machine cost for component type-2 of Technology 1.
o   Placement machine cost for component type-3 of Technology 1.
o   Placement machine cost for component type-1 of Technology 2.
o   Placement machine cost for component type-2 of Technology 2.
o   Placement machine cost for component type-3 of Technology 2.

- **Level 3**: This category includes KTVs which refer to Technology 1 on side 1 of the substrate or Technology 1 on side 2 of the substrate, independent of component types. Level 3 KTVs also refer to Technology 2 on side 1 of the substrate or Technology 2 on side 2 of the substrate, independent of components types. Examples include:
  o   Cost of substrate preparation equipment for Technology 1 on Side 1.
  o   Cost of substrate preparation equipment for Technology 1 on Side 2.
  o   Cost of substrate preparation equipment for Technology 2 on Side 1.
  o   Cost of substrate preparation equipment for Technology 2 on Side 2.

The differentiation of two assembly technologies using sample KTVs is presented in Table 5.11

Table 5.11 Differentiation of electronics assembly technologies using sample KTVs[11]

Assembly Technology →	PTH	SMT-SC-A	SMT-SC-B	SMT-SC-C
Technology 1	PTH	SMT-SC	PTH	PTH
Technology 2	-	-	SMT-SC	SMT-SC
Level 1 KTVs				
Availability of CAD Tools	Very High	High	Some	High
Level 2 KTVs				
Technology 1				
Cost of placement equipment for Component Type 1	$140,000	$300,000	$140,000	$140,000
Cost of placement equipment for Component Type 2	$120,000	$200,000-$500,000	$120,000	$120,000
Cost of placement equipment for Component Type 3	$120,000	0	$120,000	$120,000
Technology 2				
Cost of placement	0	0	$300,000	$300,000

[11] The values quoted here are for indicative purposes only.

Assembly Technology →	PTH	SMT-SC-A	SMT-SC-B	SMT-SC-C
equipment for Component Type 1				
Cost of placement equipment for Component Type 2	0	0	$200,000-$500,000	$200,000-$500,000
Cost of placement equipment for Component Type 3	0	0	0	0
Level 3 KTVs				
Technology 1				
Cost of Substrate Preparation Equipment on Side 1	0	$200,000	0	0
Cost of Substrate Preparation Equipment on Side 2	-	-	-	-
Technology 2				
Cost of Substrate Preparation Equipment on Side 1	-	-	$10,000-$25,000	$200,000
Cost of Substrate Preparation Equipment on Side 2	-	-	$200,000	-

An Example of the Use of the Technology Model

In this example, the manager is supported in identifying and understanding how the technological variables impact the MOMP "True Yield". "True Yield" is a measure of 1st Pass Yield of PCBs from an assembly process, whilst 1st Pass Yield itself is a measure of the quality of the assembly process. 1st Pass Yield represents the number of finished printed circuit boards (PCBs) that do not require rework or repair of any kind. This yield figure is determined by the yield of the technology processes (i.e. technology 1 and technology 2) used to assemble the PCBs.

As already outlined, the technology model views an assembly technology to be a mix of two primary technologies: Technology 1 and Technology 2 respectively. When the assembly technology consists of a single primary technology only, the label technology 2 becomes redundant. As an example, the 1st Pass Yield for PCBs assembled using SMT-SC-C is evaluated by computing the yield of both the PTH (Technology 1) process and the SMT-SC (Technology 2) process. The 1st Pass Yield for PCBs assembled by SMT-SC-A, is determined by evaluating the yield for the

SMT-SC (Technology 1) process, the SMT-SC-A assembly technology, consists of only one primary technology, SMT-SC.

An increase in the 1st pass yield for Technology 1 and Technology 2 results in an increase in the overall 1st Pass Yield. The 1st pass yield for each primary technology is determined by the quality of its processing on side 1 and side 2 of the substrate, or the yield for each side of the substrate. An increase in the yield achieved by a primary technology for side 1 and/or side 2 of a substrate, results in an increase in the overall yield for that primary technology.

The yield achieved by a primary technology on a particular side of the substrate, (e.g. Side 1 yield for Technology 1) is dependent on the yield of each of the component types which represent that primary technology. Thus, the yield for a primary technology on a particular side of a substrate is determined by the quality of the assembly process with respect to each component type, due to varying processes, tolerances and reliabilities etc.

The performance of any technology's placement equipment with respect to quality of placement, is a function of the capability of the machine and the potential for error during the placement process. Even though the placement equipment for the type 1 component of a primary technology may be very capable, say in terms of accuracy, it does necessarily mean that the technology performs better with respect to quality in placement, than a less accurate machine for type 1 components in another technology. If accuracy is not critical, then the level of quality reached by a less accurate machine in technology 2, may be just as high as that achieved by an accurate machine in technology 1. For example, chip on board (COB) placement equipment is very accurate. However, the potential for error during placement is high, because of the difficulty of differentiating between different COB components giving rise to the possibility of an incorrect component being placed on the board. Meanwhile, plated through hole (PTH) placement equipment is not as accurate, but the potential for error during placement is lower, because the components are easier to differentiate. One cannot say that the quality of placement is determined by placement accuracy or machine capability alone. It is a function of both machine capability and the potential for error in the placement process. Decomposition of the measure of manufacturing performance (MOMP), "True Yield", can be traced to associated key technology variables (KTVs), as indicated in Figure 5.15 shown earlier.

5.5   A Generic Programme Model for Concurrent Engineering

Concurrent engineering or concurrent new product development (CNPD) is gradually becoming the norm for developing and introducing new products to the market place.   – Haque 2003

If an organisation is to survive and prosper in today's very competitive manufacturing environment, then it must exploit market opportunities as much as possible. To do this, manufacturing organisations are increasingly migrating from the traditional *sequential product development* approach to a *concurrent engineering* approach. The goal of such a migration is to achieve significant improvements in the product development process. A by-product of this migration is a greater understanding of the factors necessary for the efficient and effective management of the product development process. These factors are based on the drivers of organisation and the management of product development, the use of appropriate technologies and techniques, the increasing number of practical experiences and knowledge of the differences of Concurrent and Sequential Engineering. By combining these different aspects of the product development work, several issues have emerged which are regarded as essential for a successful Concurrent Engineering implementation, some of which are classified as follows:

- Concurrent engineering principles.
- Concurrent engineering objectives.
- Concurrent engineering characteristics.
- Key variables in the concurrent engineering concept.
- Concurrent engineering implementation prescriptions.
- Concurrent engineering versus sequential engineering.

5.5.1 The Structure of the Programme Model

The programme model uses the following mechanisms to represent its contents and support its analysis:
- Measures of Manufacturing Performance (MOMPs)
- Key Programme Variables (KPVs).

Measures of Manufacturing Performance (MOMPs)

To measure how changes in the programme impact the manufacturing strategy; it is necessary to link the model of concurrent engineering to the business strategy via the specific MOMP map (as derived in Stage I of the AMBIT approach). This link is established to ensure that the programme framework is compatible with the manufacturing strategy framework. Because concurrent engineering is more concerned about the product development process, the main business processes which are affected by the introduction of a CE programme are the design co-ordination and the co-engineering processes.

Key Programme Variables (KPVs)

Many of the issues, which are generally considered essential to a successful implementation of Concurrent Engineering (CE), are also of equal importance to Sequential Engineering (SE). Thus, whilst such issues facilitate the overall product development process (irrespective of the approach adopted) they do not help to differentiate between CE and SE. For the purpose of the programme model, concurrent and sequential engineering parameters are identified. From these parameters, the associated Key Programme Variables are determined. A number of KPVs and associated product development parameters are shown in Table 5.12 below.

For the purpose of the programme model, key programme variables (KPVs) are defined to be essential factors of the product development process, the attributes of which assist in differentiating between Concurrent Engineering and Sequential Engineering. An example KPV is the "Problem solving approach" adopted within the organisation during product development. The attribute associated with Concurrent Engineering is "Integrated", whilst that associated with Sequential Engineering is "Localised".

Table 5.12 KPVs and product development parameters

Key Programme Variable ↓	Approaches	
	Concurrent Engineering	**Sequential Engineering**
Organisation of work	Multi-functional teams	Functional specialist
Product development process	Overlapping and continuous	Sequential
Style of project leader	Heavy-weight	Light-weight
Type of communication	Co-operative and open	Antagonistic and closed
Supplier relationship	Partnership and long term	Adversarial and short-term
Physical proximity of product developers	Co-located	Dispersed
Problem solving approach	Integrated and informal	Localised and formalised
Organisational structure	Matrix and Project based	Hierarchical and functional
Consideration of product development constraints	As early as possible	"Over-the-wall"
Product development focus	Customer orientated	Technology driven
Project management method	Integrated	Functional

Whilst Table 5.12 presents some of the KPVs and their associated attributes, two of these KPVs (namely Organisation of Work and Product Development Focus) and their respective attributes (within the context of concurrent engineering) are expanded and then categorised according to the associated macro measure(s) of competitive performance in the design co-ordination process (see Tables 5.13 and 5.14).

Table 5.13 Organisation of work

Organisation of work, KPV Attributes (within the context of CE):	

Matrix and project based multi-functional teams
- Multi-functional teams
- Matrix and project based

Organisation of work, Competitive Dimensions in the Design Co-Ordination Process:

Cost	In the best case scenario, engineering, manufacturing, marketing and purchasing work together from the outset to anticipate problems and bottlenecks and eliminate them at an early stage, in order to avoid delays in bringing the product to market and costly failures in service. Development cost is reduced indirectly by associating a money value with the reduced time-to-market [Cleetus 1993].
Quality	Careful analysis and understanding of the production and assembly process by all domain experts allows for manufacturable designs and enables the prediction of the performance of the product [Shenas 1994]. Multi-functional teams comprise manufacturing and maintenance engineers who are kept aware of the design as it unfolds and can provide suggestions that convey the constraints of the downstream perspectives to the upstream designer team members, thereby improving product quality [Cleetus 1993].
Flexibility	A multi-functional team
Time	The purpose of the multi-functional team is to open the communication channels between participating functions, so that inputs from all functions are obtained before a design is finalised [Trygg 1993]. Development time is reduced because the communication from one team member to another has no barrier of distance or hierarchical access protocols to impede information flow and work requests [Cleetus 1993]. The primary advantage is that changes can be made to the product design before it enters production, thus reducing the number of time-consuming engineering changes during full production [Trygg 1993].
Environmental Impact	The promotion of effective and efficient information flow ensures that timely and accurate information, relating to the environment can be delivered to the product development teams so that life cycle design costs can be incorporated in the overall product design process.

Table 5.14 Product development focus

Product development focus, KPV Attributes (within the context of CE):

Customer orientated
- Customer requirements
- Feedback from Market-surveys on existing products
- Use of QFD (Quality Function Deployment)

Product development focus, Competitive Dimensions in the Design Co-Ordination Process:

Cost	Using QFD a Japanese car manufacturer was able to reduce their start-up costs by 20% in 1977, 38% in 1978, and 61% in 1984 as compared to their performance prior to using QFD [Hauser *et al.* 1988].
Quality	Involvement of the customers in the development process allows the development of products that satisfy the customers' needs and desires [Trygg 1993]. Quality is improved as there is a focus on customer needs which are propagated throughout the decision making process [Cleetus 1993]. The use of QFD as a design tool brings quality control to product development. It ensures that the product meets the customer's requirements when it goes into production by a systematic focus on these customer attributes during the whole development process [Trygg 1993].
Flexibility	CE's focus on multi-functional teams and an efficient and effective information flow, encourages an integrated, dynamic organisational structure, within which product development teams can operate creatively and respond flexibly to mutable customer requirements.
Time	One of the most commonly reported advantages of using QFD (customer focus) is that it reduces the number of changes as a design enters production, in some cases up to 50%, consequently decreasing the time needed to get a design into production [Trygg 1993].
Environmental Impact	Life cycle design is a large part of the product development process. It is directed at the consideration of the total costs of a product over its entire life, to the manufacturer (i.e. the company), the user and society, and attempts to minimise such costs [Alting 1991]. By taking consideration of the life cycle design costs, more environmentally 'benign' products can be produced. The following table shows the costs incurred by the company, the user and society from a product distribution and a product disposal perspective.

	Company Cost	**User Cost**	**Society Cost**
Product Distribution	Transportation, storage and waste	Transportation and storage	Waste, pollution, packaging and health damages
Product Disposal	–	Disposal dues	Waste handling, disposal, health damages and pollution

5.6   The Specific Models

The generation of a specific model occurs with the user customisation of the generic technology or programme model (Figure 5.16). As with the customisation of the generic MOMP map in Stage I of the AMBIT approach, customisation of the generic technology or programme model is executed following the user's examination of the generic technology/programme model, and identification of 'shortcomings' in the model from the point of view of the particular organisation or situation under study. Customisation offers the user a chance to edit the generic technology or programme model, in order to add, modify or remove key technology variables (KTVs) and/or key programme variables (KPVs), which are regarded as inappropriate to the organisation under study. This transition from a generic model to a specific model reflects the movement along the continuum of generality; from a general model to partial model, and then following instantiation, from a partial model to a particular model.

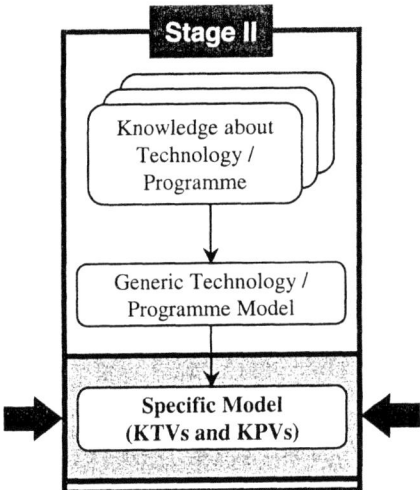

Figure 5.16 Progress in Stage II – the specific model

As an example, in the case of the technology model, the range of values for each of the key technology variables (KTVs) (i.e. lead pitch, component density, board silicon density, number of component leads, component size, package presence and board wiring density) is very wide. Table 5.10 presented in section 5.4.1 earlier provided sample ranges of the KTVs for each of the following popular technologies i.e. Plated Through Hole, (PTH), Surface Mount Technology (SMT), Fine Pitch Technology (FPT), Chip On Board (COB) technology, Multichip Module (MCM) technology. The precise values and their relevance to the products and manufacturing system under analysis have to be captured. Looking back to the list of KTVs for electronics assembly presented in Table 5.3 (section 5.4) earlier, it is clear that very many of these KTVs will have values particular to an individual enterprise. For example, the cost values (component costs, substrate costs, equipment cost) will depend on many factors including location, tax regime, relationship with suppliers etc. The set up and process times as well as the repair times are somewhat dependant on the capability of the manufacturing process and skilled operators. Each KTV will need to be considered, its relevance to the situation under review considered and particular values assigned as appropriate.

A customised generic technology/programme model is called a specific model. Once the user is satisfied that the specific model appropriately reflects the situation under study, the models can be analysed using a set of analysis tools. These tools (see chapter 6) support the evaluation of the strategic level consequences of changes in interconnect technology and concurrent engineering. The development of such tools and the subsequent analysis of the models are covered in Stage III of the AMBIT approach, which will be discussed in Chapter 6 following.

5.7   Conclusion

During Stage II of the AMBIT approach, a bridge is established between the strategic management manufacturing and manufacturing domains. This is important in order to help management assess the strategic implications of changes in either the technology or the programme. The development of models in this Stage was elucidated using the examples of interconnect technology and concurrent engineering.

In Stage II of the AMBIT approach, knowledge about the technology and programme is captured and redefined in the form of generic technology and programme models. The structure of these models is modular in nature in order to both deal with the complexity of the technology and programme domains and to facilitate user customisation. Whilst the technology and programme knowledge is derived through a series of interviews and literature reviews, care is taken such that the level of information detail presented to the targeted user is not too technically dense. If the AMBIT tool is developed to support decisions concerning the introduction or modification of existing technologies and/or programme, and act as a learning tool, it is vital that the information presented to the targeted user(s) be requisite to their needs.

Key technology variables (KTVs) represent key product and/or process features which differentiate, for example, one interconnect technology class from another. Key programme variables (KPVs) represent the essential features of the product development process; features whose parameters differentiate between, for example Concurrent and Sequential engineering. A KTV and a KPV can be viewed as more operational differentiators than strategic performance indicators (SPIs) or measures of manufacturing performance (MOMPs). Because a KTV/KPV is a key characteristic of any technology/programme class, a value can be attributed to it. Indeed, given the structure of the technology and programme models, each KTV/KPV influence one or more of the business processes, and may also impact one or more of the five measures of competitive performance within each process. Furthermore, based on the use of MOMPs as 'connectors' between the technology/programme and the management models, the repercussion of changes in a technology/programme on the business strategy of the organisation can thus be understood.

Chapter 6

The AMBIT Approach – Stage III

Keywords: PROTIMA (Programme/Technology Implementation Analyser), KASPER, (knowledge based advisor and strategic programme evaluator), causal effect, AHP (Analytic Hierarchy Process), pair-wise comparisons, analysis of specific MOMP map, analysis of specific model, CBR (case-based reasoning), retrieve, reuse, revise, retain.

Chapter Objectives: The purpose of this chapter is to introduce some of the analysis tools which can be used to support the manager or team of managers in evaluating the strategic effects of a particular technology and/or programme implementation and/or change. The tools used to illustrate the possible analysis approaches are; Analytic Hierarchy Process (AHP), Case Based Reasoning (CBR) and Semantic Modelling. Having read this chapter, the reader should become familiar with:

- The PROMTIMA analysis tool, which comprises KASPER and AHP. This tool can be used to reason between the specific technology or programme model and the specific MOMP map.
- Case Based Reasoning (CBR) and how it might be used.

6.1 Introduction

Lissack and Roos (2001, p. 57) point out that decision making processes and models are often based on four false assumptions, the world is stable enough that changes that may occur are foreseeable; prediction is possible; boundaries are clearly defined; and outcomes are more important than processes. As a consequence of the above assumptions, decision makers often fail to create, develop, or exploit opportunities.   – Pech *et al.* 2003

Strategic management is the process of assessing an organisation and its competitive environment, in order to identify whether or not the organisation can satisfy its long-term objectives and avail of market opportunities [Stoney 2001, Alkhafaji 1995]. | Thus, strategic management requires lucid thinking, a clear purpose and the commitment of others to win [Lawrie 1996]. In supporting the effectiveness of the strategic management task, strategy managers require tools, which help them assess the strategic impact of new technology and/or programme implementation on their organisation. Clearly, the question of whether or not to implement a new (or modify an existing) technology and/or programme is far more likely to be answered positively, if the strategic impact of such a decision can be quantified.

Chapter 4 described Stage I of the AMBIT approach. Stage I concerns the linkage of the business strategy - (expressed in terms of Critical Success Factors (CSFs) and Strategic Performance Indicators (SPIs) - to the manufacturing strategy - expressed in terms of specific Measures Of Manufacturing Performance (MOMPs).

Chapter 5 described Stage II of the AMBIT approach, which represents the linkage of the more operational elements of the manufacturing strategy to the Key Technology Variables (KTVs) and Key Programme Variables (KPVs). See Figure 6.1 below.

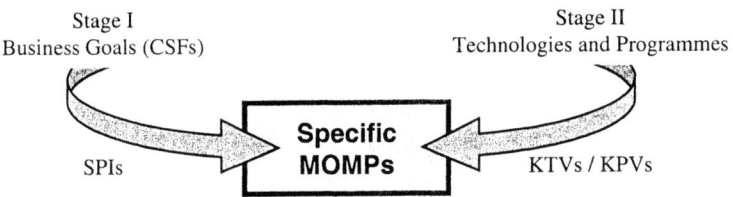

Figure 6.1 Linking the business goals to the technologies and programmes via the specific MOMPs

Finally, in Stage III (the subject matter of this chapter), some analysis tools are described that can be used to support the manager in evaluating the strategic effect/s of implementing or modifying a particular technology or programme. Therefore, the purpose of Stage III (Figure 6.2) is to analyse the technology and programme models developed in Stage II. During this analysis, the strategic influence of the technology and programme is evaluated, alternative approaches are assessed and a particular course of action is selected. Key analysis tools that are employed in Stage III to fulfil these goals are:

- The PROTIMA (Programme/Technology Implementation Analyser) tool (Section 6.2).
- A Case Based Reasoning (CBR) tool (Section 6.3).

It should be noted that such tools are intended to be used only when both the specific technology model and the specific MOMP map have been validated and are in a stable state. It is also important to emphasise that the tools presented here are examples to illustrate the possible analysis approaches.

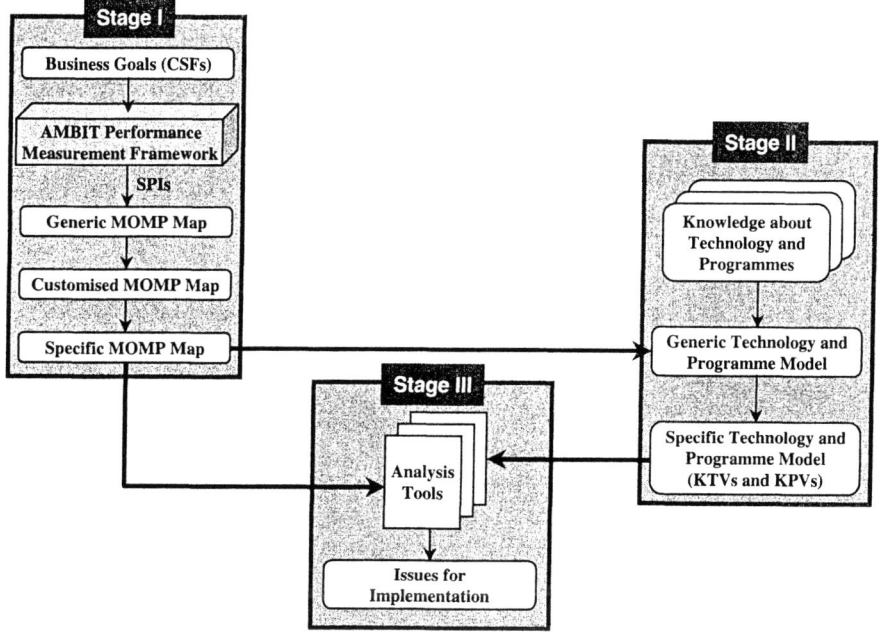

Figure 6.2 The AMBIT approach – Stage III

A complex approach such as AMBIT cannot be fully exploited unless it is adequately supported by computer-based tools. During the development of the AMBIT approach, the authors researched and developed a number of laboratory prototype tools to articulate and validate the ideas underlying this approach. Throughout this chapter reference will be made to these tools that help explain the underlying ideas. The screen-shots presented in this chapter are taken from these prototypes.

6.2 The PROTIMA Tool

The PROTIMA (Programme/Technology Implementation Analyser) analysis tool reasons between the specific model and the specific MOMP map. The PROTIMA tool adopts a dual approach (Figure 6.3), by using:

- AHP (Analytic Hierarchy Process) approach, as developed by Saaty [1980], is used to prioritise the elements within the specific MOMP map hierarchy. It therefore reasons across the specific MOMP map.
- KASPER (Knowledge Based Advisor and Strategic Programme Evaluator [Jackson, 1991]) supports the strategy manager in assessing the influence of a new manufacturing technology and/or programme, or changes to an existent

technology and/or programme, on the specific MOMP map. KASPER reasons across the specific technology and MOMP models.

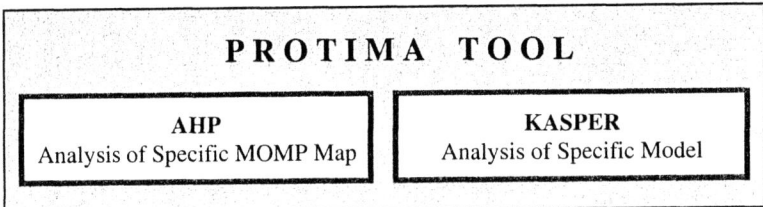

Figure 6.3 The PROTIMA tool

6.2.1 AHP – Analysis of Specific MOMP Map

Although analytic hierarchy process (AHP) is effective to use for management decision making, it can be defective if used improperly.   – Cheng *et al.* 2002

The Analytic Hierarchy Process[12] (AHP) approach, as developed by Saaty [Saaty 1980], is used to prioritise the elements within the specific MOMP map hierarchy. It therefore reasons across the specific MOMP map. AHP provides a theory and a corresponding methodology, which support the modelling of unstructured problems. AHP is used to select from a set of alternatives once the decision process is structured in a hierarchical form. Udo states that the main advantages accruing through the use of the AHP technique is its simplicity and its thoroughness in handling difficult real-life problems. "It provides greater utility in applications where information is either incomplete or not available" [Udo 2000].

The evaluation approach used by AHP is the pairwise comparison technique. In pairwise comparisons, all of the entities on a level are compared with each other, with respect to their influence on one of the entities in the next highest level. On completion of the pairwise comparisons, the alternatives on the lowest level are weighted with respect to their importance in achieving the overall goal. In AHP a relative scale, based on a series of verbal judgements is used (Table 6.1).

Table 6.1 Scale for pairwise comparisons

Verbal Judgement	Numerical Rating
Absolutely more important	9
Very strongly more important	7
Strongly more important	5
Moderately more important	3
Equally Important	1

---

[12] Weber [1993] has developed a tool called Automan, which allows the user to carry out AHP analysis of any problem.

The questions take the form of: "with respect to <entity at the next highest level> how much more strongly is <first entity> more important than <second entity>"? For example, a level has four different entities A, B, C and D (Table 6.1), which are pairwise compared. If:

- Activity A is strongly more important than activity B, the rating is 5.
- Activity A is between strongly more important and very strongly more important than activity C, the rating is 6.
- Activity A is very strongly more important than activity D, then the rating is 7.
- Activity B between moderately more important and strongly more important than activity C, the rating is 4.
- Activity B is between strongly more important and very strongly more important than activity C, the rating is 6.
- Activity C between moderately more important and strongly more important than activity D, the rating is 4.

The resultant judgements are represented in a matrix format as follows (Table 6.2):

Table 6.2 Pairwise comparison matrix

	A	B	C	D
A	1	5	6	7
B	.2	1	4	6
C	.167	.25	1	4
D	.143	.167	.25	1

In the interest of symmetry, when activity A is 5 times activity B (i.e. strongly more important), activity B is thus 0.2 times activity A. 1's are used in the diagonal because logically, something cannot be more important than itself. For perfect consistency, if activity A is 2 times more important than activity B, and activity B is 3 times more important than activity C, then activity A should be 6 times more important that activity C. However, if this not the case, inconsistency is said to occur.

After defining the pairwise comparison matrix, the weightings for each of the criteria are calculated by normalising the eigenvector, which has the largest eigenvalue of this matrix. There are several ways by which this can be done. Saatay [Saatay 1980] proposes the following three methods:

Method 1:
1. Sum the elements in each row,
2. Normalise, by dividing each sum by the total of all sums, so that the results equal to 1,
3. The first entry of the resulting vector is the priority of the first activity, the second of the second activity etc.

Method 2:
1. Divide the elements of each column by the sum of that column and then add the elements in each resulting row,
2. Divide this sum by the number of elements in the row,
3. This represents the process of averaging over the normalised columns.

Method 3: (used in this example)
1. Take the sum of the elements in each column of the pairwise comparison matrix (1.51, 6.417, 11.25, 18),
2. Form the reciprocals of these sums (0.662, 0.156, 0.089, 0.056),
3. Calculate the weights (i.e. the estimated solution vector) by dividing each reciprocal by the sum of the reciprocals, as shown below:

$$A = \frac{0.662}{0.963} = 0.687$$
$$B = \frac{0.156}{0.963} = 0.162$$
$$C = \frac{0.089}{0.963} = 0.092$$
$$D = \frac{0.056}{0.963} = 0.058$$

If the matrix is consistent, the variance across the weights should be relatively small. However, as this is not the case in the example under discussion, the inconsistency of the matrix (i.e. the consistency index) is quantified by evaluating the following expression:

$$\frac{\lambda_{max} - n}{n - 1}$$

where λ_{max} is the largest eigenvalue and n is the number of elements in the vector, and it is calculated by:

1. Multiplying the original matrix of judgements by the estimated solution vector:

$$\begin{bmatrix} 1 & 5 & 6 & 7 \\ 0.2 & 1 & 4 & 6 \\ 0.167 & 0.25 & 1 & 4 \\ 0.143 & 0.167 & 0.25 & 1 \end{bmatrix} \times \begin{bmatrix} 0.687 \\ 0.162 \\ 0.092 \\ 0.058 \end{bmatrix} = \begin{bmatrix} 2.455 \\ 1.015 \\ 0.479 \\ 0.206 \end{bmatrix}$$

2. Dividing the first component of this new vector by the first component of the estimated solution vector, and the second component of this vector by the second component of the solution vector, etc. i.e.

$$\begin{bmatrix} 2.455 & 1.015 & 0.479 & 0.206 \\ 0.687 & 0.162 & 0.092 & 0.058 \end{bmatrix}^{\mathsf{T}} = \begin{bmatrix} 3.57 \\ 6.26 \\ 5.206 \\ 3.55 \end{bmatrix}$$

3. Divide the sum of the components (18.586) by n (n = 4). This yields 4.646.
4. Hence, the value of λ_{max} is 4.646.

λ_{max} is also called the principal eigenvalue. The closer λ_{max} is to n (in this case 4), the more consistent is the result. In the event that a large number of pairwise comparisons have to be made, it is not realistic to demand perfectly consistent matrices. However, it is still important that the matrices be as consistent as possible. In this example, the consistency index (CI) is 0.2153 ((4.646 – 4) / 3). In order to identify whether or not the deviation from consistency is acceptable, the consistency ratio (CR) is evaluated, by dividing the value for the CI by the average random index (RI) value. This value is obtained from the table showing the average consistency of random matrices (dependent on the number of entities) (Table 6.3). If the resultant CR is 0.10 or less, then the matrix of judgements is deemed acceptable.

In this example, the consistency ratio (CR) of 0.2392 (i.e. 0.2153/0.90) is outside the acceptability tolerance of 0.10 or less. Thus, the decision maker will need to reassess the initial matrix of judgements. The calculation of the final rating of each of the alternatives at the lowest level is obtained by following each path upward to the objective, multiplying the weightings in the process. The sum of all the paths leading from an alternative is the final rating for that alternative.

Table 6.3 The random index table

1	0	6	1.24	11	1.51
2	0	7	1.32	12	1.48
3	0.58	8	1.41	13	1.56
4	0.9	9	1.45	14	1.57
5	1.12	10	1.49	15	1.59

Using AHP to Analyse the Specific MOMP Map

In PROTIMA, the MOMPs are not rated as alternatives to each other, but are instead prioritised in terms of their importance to the organisation's goal. In accomplishing this, the overall goal (expressed as the success of the company) is broken down into sub-goals, which are rated in terms of their importance to this goal. This decomposition process, where the parent goals are broken down into child goals and the child goals rated in terms of their importance to the parent goal continues until the specific MOMPs are reached. This procedure reflects the decomposition of the goal (i.e. success of the company) into strategic performance indicators (SPIs), which are further decomposed into MOMPs and then to specific MOMPs.

In the following example, cost of manufacturing and quality of supply, which represent the SPIs, are vital to the success of the company. The decomposition of these SPIs into intermediate level and lower level MOMPs is shown in Figure 6.4.

On validation of the hierarchy, the elements are prioritised using the pairwise comparison approach. Figure 6.5 shows the specific MOMP map of Figure 6.4 after prioritisation.

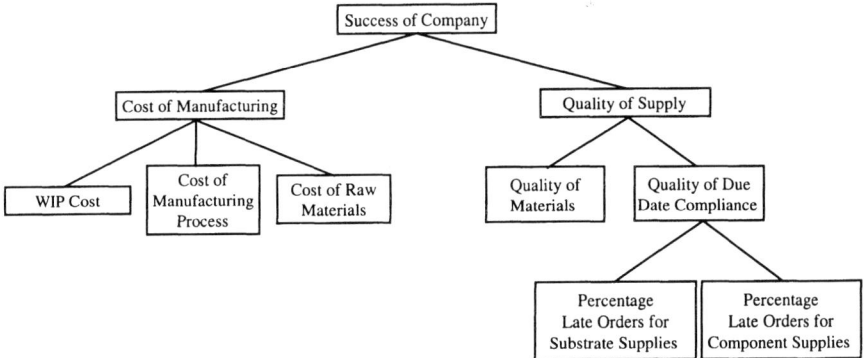

Figure 6.4 Hierarchy of SPI decomposition to specific MOMPs

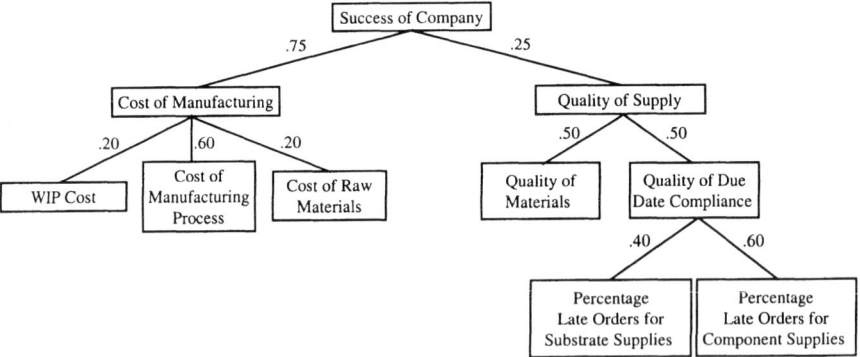

Figure 6.5 The sample specific MOMP map after prioritisation

After prioritisation, the user calculates the priority of the lower level MOMPs with respect to the elements on the higher levels. For example, the priority of the percentage late orders for component supplies, with respect to success of the company is 0.6 x 0.50 x 0.25 = .075.

The other MOMPs are prioritised in a similar manner. Thus:

- Importance of WIP cost with respect to the success of the company is 0.2 x 0.75 = 0.15.

- Importance of cost of manufacturing process with respect to the success of the company is 0.6 x 0.75 = 0.45.
- Importance of cost of raw materials with respect to the success of the company is 0.2 x 0.75 = 0.15.
- Importance of quality of materials with respect to the success of the company is 0.5 x 0.25 = 0.125
- Importance of percentage late orders for substrate supplies with respect to the success of the company is 0.4 x 0.5 x 0.25 = 0.05.
- Importance of percentage late orders for component supplies with respect to the success of the company is 0.6 x 0.5 x 0.25 = 0.075.

The sum of all these values (0.15 + 0.45 + 0.15 + 0.125 + 0.05 + 0.075) should equate to 1.0. From this data one can clearly notice that the cost of manufacturing process is vital to the success of the company.

The AHP approach is useful when a decision must be made between mutually exclusive alternatives. Its ability to deal with flawed and inconsistent data (however, there must be a reasonable level of consistency across all of the judgements) represents one of its greatest strengths. Indeed, this characteristic is of particular use when dealing with an area such as strategic decision making where many of the judgements are somewhat subjective and uncertain.

Mintzberg suggests that most successful managers tend to use intuition and hunches in decision making. Indeed, he believes that managerial decision making processes seem to be more relational and holistic than ordered and sequential, and more intuitive than intellectual [Mintzberg 1976]. As a result, decision making involves both qualitative and quantitative data. Therefore we can suggest that when managers are making decisions, a structured model of the domain in question is more important than a structured model of the decision making process. By browsing a model of the problem domain, the manager can gain a greater understanding of and insight into the problem. His holistic processes are supported, and as a result, a more informed decision can be made. However, most Decision Support Systems (DSSs) concentrate on modelling the decision making process, as opposed to modelling the problem domain. Furthermore, since decision making in the strategy formulation domain is often fraught with uncertainty, the manager should be able to express a degree of belief or confidence in his judgement and also be able to prioritise the specific MOMPs with respect to their importance to the overall goal. We believe that the PROTIMA tool (which comprises KASPER and AHP) fulfils these needs.

6.2.2 KASPER – Analysis of Specific Models

KASPER (Knowledge Based Advisor and Strategic Programme Evaluator). KASPER supports the strategy manager in assessing the influence of a new manufacturing technology and/or programme, or changes to an existent technology and/or programme, on the specific MOMP map. KASPER reasons across the specific model

KASPER is a knowledge-based advisor and strategic programme evaluator. It allows the subjective view of an individual or group of individuals concerning strategic issues, to be encoded in a knowledge base. This knowledge base is specifically designed to cater for qualitative and quantitative knowledge. Provision should also be made in the knowledge base for the strategy manager to express his level of confidence in the judgements he makes.

KASPER captures unstructured and intuitive knowledge by allowing qualitative relationships to be expressed in both graphical and linguistic form. This modelling of qualitative knowledge is extremely important, as strategic decision making often involves making decisions based on highly unstructured and often conflicting information. Such decisions are often based as much upon intuition and experience, as rationality and logic. As a result of this uncertainty of qualitative knowledge and the unstructured nature of the strategic decision making environment, knowledge representation using traditional knowledge base techniques is frequently restricted and many analytical techniques and tools tend to disregard the intuitive aspect of strategic decision making. KASPER however, emphasises and supports intuitive thought and problem solving, in addition to the more formal methods of analysis. .

The loose structure of the KASPER knowledge base encourages intuitive thought, and allows knowledge which exists only in the form of experiences or hunches to co-exist with more formal knowledge. In supporting the encoding and analysis of such qualitative and quantitative knowledge, KASPER architecture comprises of several distinct elements. Figure 6.6 depicts the interaction among the elements.

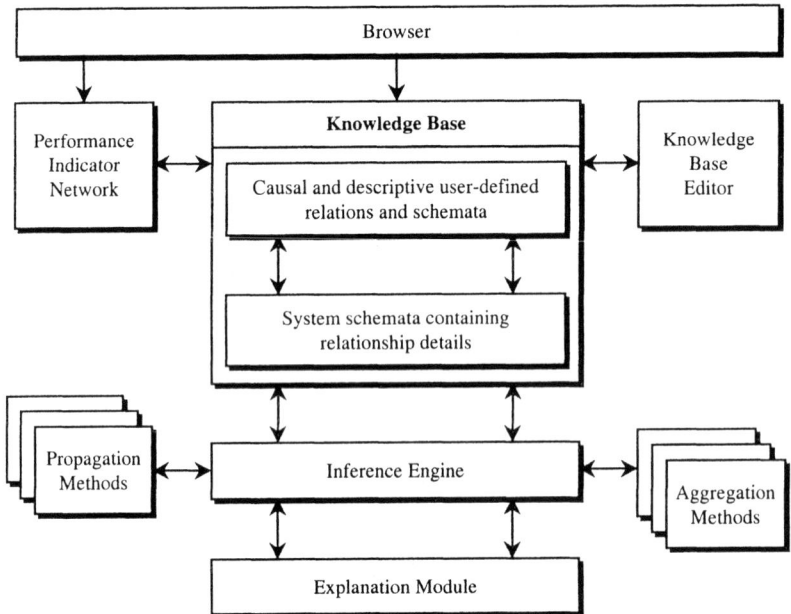

Figure 6.6 Overview of the KASPER architecture

The functionalities of the key KASPAR elements are:

- A browser. The browser supports ease of modification of the knowledge base, so that the strategy manager can experiment with a range of values.
- A knowledge base. The qualitative and quantitative knowledge is stored in the knowledge base.
- An inference engine. The inference engine performs aggregation and propagation of qualitative relationships and their associated degrees of belief throughout the knowledge base.
- An explanation module. The explanation module provides an explanation of its conclusions, so that the strategy manager can interrogate the system in order to determine how its conclusions were generated.

At the core of the inference engine (and indeed qualitative reasoning) is the notion of a causal effect. A causal effect describes the effect that one object has on another. It is because of this representation mechanism in the knowledge base that the inference engine can provide a qualitative measurement of the causal effects of a new technology or programme implementation (or changes to an existent technology or programme) on a selected variable or set of variables.

Figure 6.7 shows the causal effect of process control on level of quality. Literally translated, the relationship states that the level of quality is increased by process control.

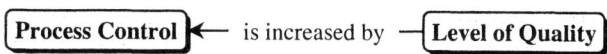

Figure 6.7 Simple relationship between level of quality and process control

Whilst the relationship in Figure 6.7 is easily understood, it does not show the magnitude of the influence which process control has on level of quality (i.e. is level of quality increased a lot or a little by process control?). To overcome this shortcoming, the extent of the increase in the level of quality for each possible increase in process control needs to be computed and stored in the system, such that a graphical representation of this information can then be drawn. This graphical representation is called the casual effect (CE) function. It is drawn over a normalised scale in order to allow concepts, which may differ significantly in absolute terms to be related together easily (e.g. process control and level of quality) (Figure 6.8). The origin (0.0) represents the relationship between two objects prior to any change taking place (in either object). 1.0 represents the maximum possible change in either object. Relationships can be defined graphically in either linear or in non-linear form.

The positive direction is defined as the direction of desirable change for a given schema. Meanwhile, the negative direction is defined as the direction of undesirable change. For example, the positive direction for level of quality shows an increase in the level of quality, which is clearly desirable. However, a negative direction for operating expenses indicates increasing costs, which represents an undesirable state. Despite the fact that the graph allows concepts to be related together easily, it can often be difficult to interpret such graphs and remember the implications of the relationships. This is particularly true if the model is used to describe a large number of inter-related, causal effects. Because of this shortcoming, both the causal effect and the graphical representation approaches should be used in the KASPER tool.

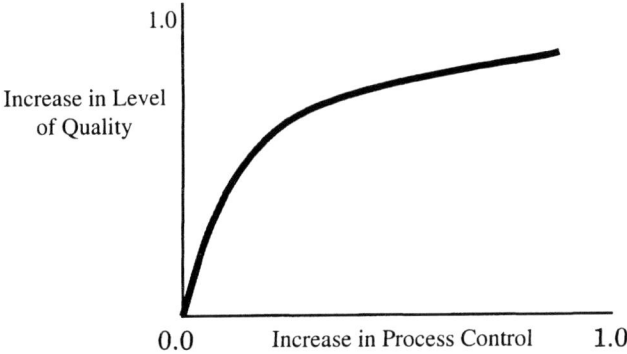

Figure 6.8 Graphical representation of the relationship between process control and level of quality

KASPER allows the user to evaluate the influence of one schema (e.g. process control) over others (e.g. level of quality) because of the following features:

- **The Causal Effect (CE) function.** The CE function graphically describes the full range of effects that one schema has on another schema (Figure 6.8).

- **The Independent Causal Effect (ICE) value.** Whilst the CE function shows the whole range of possible effects that one schema has on another, only one such effect is applicable at any given time. This particular effect at this particular time is known as the independent causal effect (ICE) value. Taken directly from the CE function, it measures the influence of one schema on another. The ICE value is represented as a numerical value having a linguistic equivalent [Jackson 1991]. As an example, if the user selects a value in the range of 0.01 to 0.19 to relate the influence of process control on level of quality, then this means that level of quality is increased very little by process control. However, because there is often an element of uncertainty associated with this judgement, it also needs to be described graphically. This is done using the confidence function (CF).

- **Degree of Confidence (DOC).** The DOC value, which is associated with each ICE value, represents the confidence level that the strategy manager has in the ICE value. The DOC value is defined on a scale from 0.0 to +1.0. As in the case with the ICE values, the DOC values are represented as numerical values having linguistic equivalents (e.g. 0.0 represents complete uncertainty, whilst +1.0 describes complete certainty). DOC values are defined graphically using a Confidence Function (CF) (Figure 6.9). The dotted line represents the point up to which DOC values are defined. Values drawn beyond this point are ignored.

- **The Confidence function (CF).** The confidence function illustrates the strategy manager's degree of certainty in the ICE value (Figure 6.9). The x-axis contains the range of ICE values, whilst the y-axis depicts the degree of confidence (DOC) in these values. Based on the ICE value, the corresponding DOC value can be easily identified.

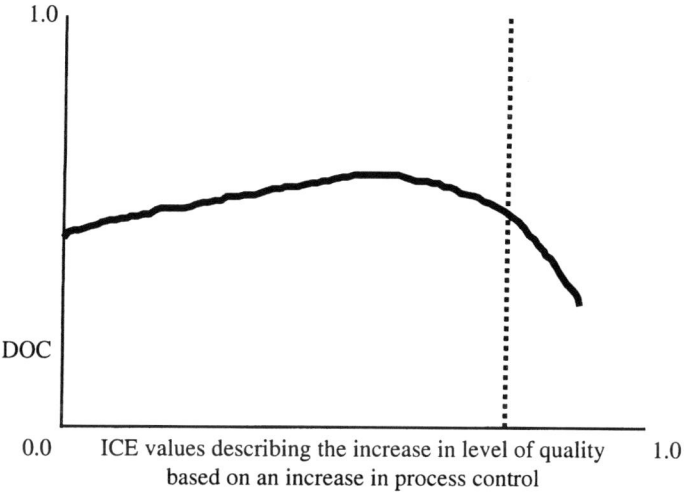

Figure 6.9 An example based on the positive values of Figure 6.8

- **The Aggregated Causal Effect (ACE) value.** Because the selected schema is often influenced by more than one schema, the net influence on the selected schema is a combination of the ICE values for each of the influencing schemata. This net influence is termed the Aggregated Causal Effect (ACE) value for the leaf schema. It has a value between -1.0 and +1.0. Negative values show an impact in the negative direction. If only one schema impacts the selected schema, then the ICE and the ACE values are similar.

- **The Aggregated Degree of Confidence (ADOC) value.** Just as the total influence of all influencing schemata on a selected schema is considered (i.e. the ACE value), the total confidence level in this value is computed. This is called the ADOC value and it is calculated by combining the different DOC values associated with each ICE value. It is measured between 0.0 and +1.0, and is an expression of the degree of confidence in the ACE value.

Using KASPER

In KAPSER the problem domain is represented as a semantic network[13] composed of schemata (objects), with relationships between schemata defined by a CE (Causal Effect) function and a corresponding confidence function. KASPER supports the strategy manager in analysing the influence that related schemata would have on the selected schema in the network. Indeed, the ACE on a particular schema is derived after first establishing the ICE value of each of its neighbouring schemata, each of which in turn will depend on other schemata. This process continues in a recursive fashion until a situation is reached for every relationship where the ICE describing a given relationship is derived not from a CE function, but from an absolute value. The ACE and ADOC values are then calculated. Because KASPER provides a means of tracing causal effects through a network, the effect of a technology and/or programme on a measure of manufacturing performance (MOMP) can be assessed.

The network segment of Figure 6.10 shows the immediate schemata which impact design complexity in the design of electronic assemblies (i.e. number of leads, number of components and board size). It also depicts the schemata which are in turn affected by design complexity (i.e. substrate preparation time, bonding time and cleaning time). In order to evaluate the effect on substrate preparation time, bonding time and cleaning time as a result of the influence of the schemata impacting design complexity, the influence of number of leads, number of components and board size on design complexity must first be propagated. Once this is accomplished, the influence of design complexity on substrate preparation time, bonding time and cleaning time can be propagated.

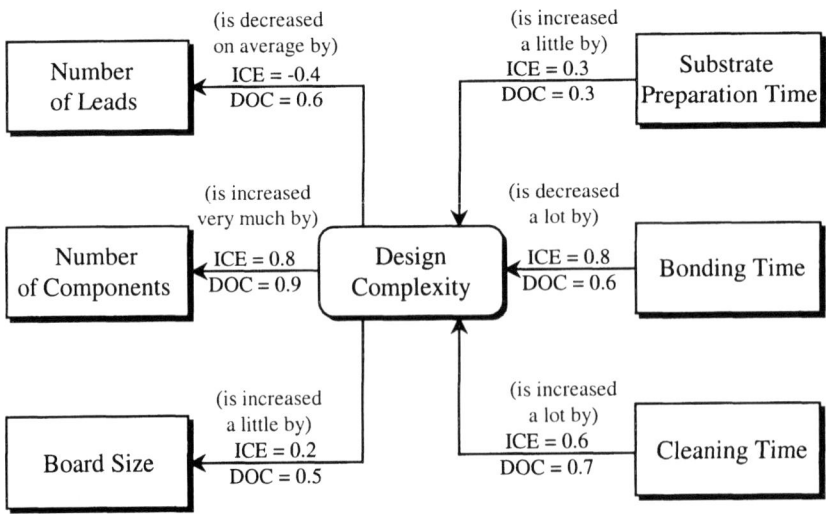

Figure 6.10 A sample network fragment showing ICE and DOC values

---

[13]A semantic network is a graphical representation of the information and its interrelationships. It is an intuitive way of presenting facts.

In this example, the ACE and ADOC values for number of leads, number of components and board size have been calculated and the ICE and DOC values of the relations connecting them to design complexity have already been propagated, based on the CE and CF for the relations. The ACE and DOC values for design complexity are aggregated using the following weighted average calculation to determine the net impact of the ICE values of number of leads, number of components and board size:

$$ACE = \frac{\sum_{i=1}^{n}(ICE)_i\,\omega_i}{\sum_{i=1}^{n}\omega_i}; \quad w_i = 100 \times \left(1 - \left(DOC_{max} - DOC_i\right)\right)$$

The resulting ACE value in this case reflects the net impact of all root schemata (i.e. number of leads, number of components and board size) on the leaf schema (i.e. design complexity), whilst the ADOC value represents the net confidence level of that effect on the leaf schema. Thus, the ADOC value is a combination of the different DOC values associated with each ICE that was used to calculate the ACE. DOC_{max} represents the highest DOC value of the different relationships impacting the root leaf schema. In this example, DOC_{max} is 0.9 (the value associated with number of components). The ICE values which have the DOC_{max} are assigned the highest weighting factor. As a result, the more certain an ICE value the higher the weighting factor it receives. ICE values with a high degree of uncertainty tend to be discounted i.e. ICE values of 0.0 and their associated DOC values are ignored in the calculation. Figure 6.11 shows the calculation of the ACE value of design complexity.

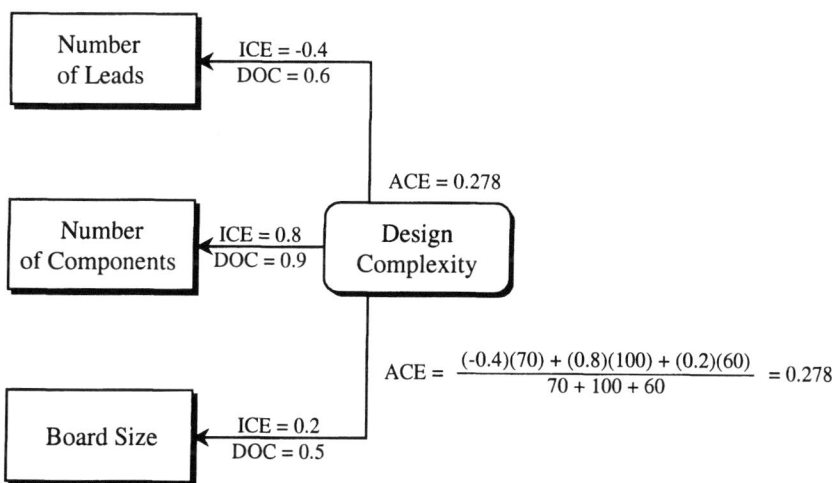

Figure 6.11 The calculation of the ACE for design complexity

The ADOC or aggregation of the DOC values is performed in a similar fashion, using the following equation;

$$ADOC = \frac{\sum_{i=1}^{n}(DOC)_i \omega_i}{\sum_{i=1}^{n}\omega_i}; \quad w_i = 100 \times \left(1 - \left(DOC_{max} - DOC_i\right)\right)$$

Figure 6.12 illustrates the ADOC calculation for the sample network fragment.

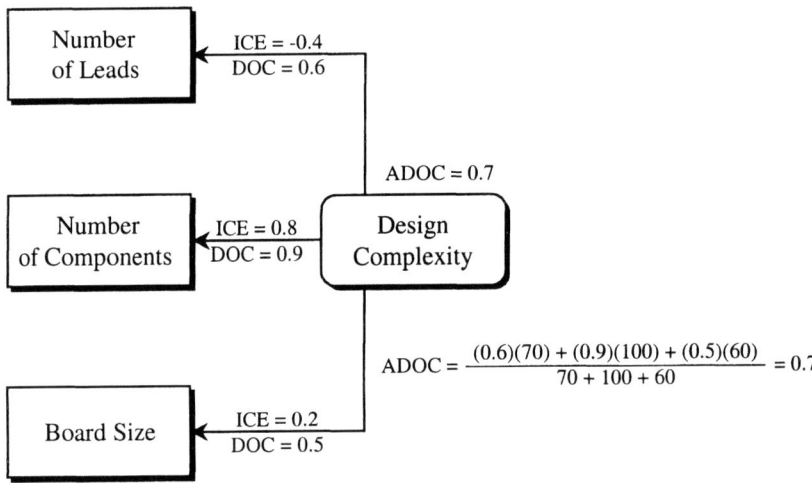

Figure 6.12 The calculation of the ADOC value for design complexity

This cycle of aggregation and propagation continues until the specific measures of manufacturing performance (MOMPs) are reached (i.e from the key technology variables (KTVs) and key programme variables (KPVs) to the Linking objects (LOs) to the specific MOMPs). When this happens, the aggregated causal effect (ACE) and aggregated degree of confidence (ADOC) values of these specific MOMPs (i.e. specific MOMPs) can be viewed, in order to determine the effects of the KTVs/KPVs on them. The ACE and ADOC values of other schemata can also be examined to see how they influence those specifics MOMPs.

It is important to stress that the goal of the analysis tools, is not necessarily to provide answers but rather to provide insights into the problem. KASPER allows managers to define and examine extremely complex and ill-defined problems in a meaningful manner. As a result, the manager's understanding and familiarity with the problem domain is enhanced and a more informed decision can be made.

6.2.3 Using the PROTIMA Tool – A Case Study

In the following example, the manner in which PROTIMA helps the manager examine the effects of introducing a new interconnect technology in an electronics organisation is described. Also, the ease with which he is supported in browsing the hierarchy, altering a single judgement and viewing the changes that propagate right through to the specific MOMPs is highlighted.

In this particular example, the organisation is involved in printed circuit board (PCB) assembly i.e. purchasing microchips, substrate etc. and producing PCB assemblies from them. The raw materials are available from a large number of suppliers at a nearly uniform price, and there is no difference between the final product of this organisation and that of its competitors since they all use standard parts to produce standard assemblies. This particular organisation is considering a change from one interconnect technology class to another i.e. from plated through hole technology (PTH), to chip on board technology (COB). Thus, the management team needs to know the impact of such a change on his strategic performance indicators (SPIs) and then ultimately on the organisation's goal.

As described in Chapter 4, the AMBIT performance measurement (PM) framework provides a means of translating the business goals of a manufacturing organisation, expressed in terms of its critical success factors (CSFs), into a specific MOMP (measures of manufacturing performance) map. Each point within the AMBIT cube (see Figure 4.4), as defined by its co-ordinates, is termed a strategic performance indicator (SPI). The strategy of a manufacturing organisation is expressed in terms of its critical success factors (CSFs) which are in turn used to identify corresponding SPIs. SPIs can provide answers to the following: how will the organisational performance be evaluated? and how is the organisation doing? Thus, an SPI is a business metric which is a recognised measure of strategic performance.

In this example, the organisation's mission statement is 'To become the domestic market leader by producing lower cost PCB assemblies than our competitors'. However, as it is difficult to relate this high level statement to the strategy pursued by manufacturing, it is easier to express it in more quantifiable terms (Figure 6.13) i.e. in the form of SPIs.

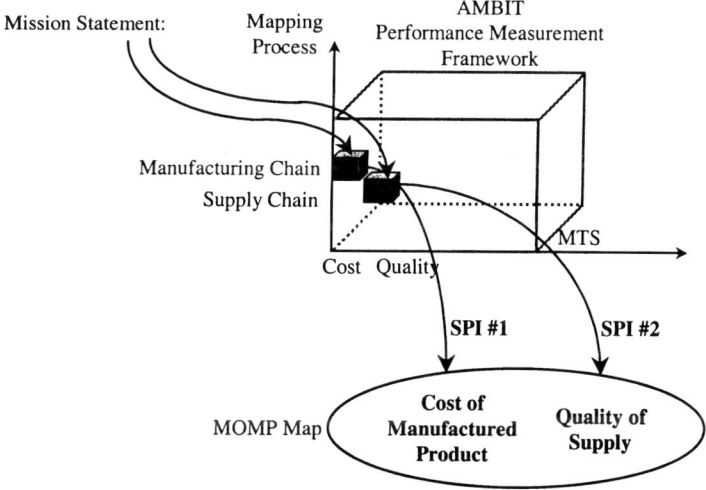

Figure 6.13 Translation of the mission statement to relevant SPIs

In Figure 6.13, the SPIs that reflect this mission statement are:
- Cost in the manufacturing chain in a make to stock environment i.e. to manufacture the product more cheaply than that of the competitors.
- Quality in the supply chain in a make to stock environment i.e. to ensure that raw materials are reliably sourced, with good supplier due date performance and low reject rates of supplied components.

Using the AMBIT performance measurement framework, these SPIs are decomposed into the following MOMPs:
- WIP cost.
- Cost of manufacturing process.
- Cost of raw materials.
- Quality of materials.
- Percentage late orders for substrate supplies.
- Percentage late orders for component supplies.

To support the strategy manager in his decision, a technology model is constructed which represents the impact that the change from plated through hole (PTH) to chip on board (COB) would have on the specific MOMPs. The specific MOMPs in this case being: WIP Cost, Cost of Manufacturing Process, Cost of Raw Materials, Quality of Materials, Percentage Late Orders for Substrate Supplies and Percentage Late Orders for Component Supplies. This technology model depicts the causal links between the implementation of COB and the specific MOMPs. For ease of analysis and clarity, the technology model is split between each individual specific MOMP. Figure 6.14 shows the network segment where COB is linked via the key technology variables (KTVs) to the Cost of Raw Materials MOMP. If no direct relationship exists between a specific MOMP and a KTV, then the specific MOMP is decomposed into linking MOMPs.

The rest of the technology model is built along similar lines, featuring causal links from COB to KTVs and from the KTVs to specific MOMPs.

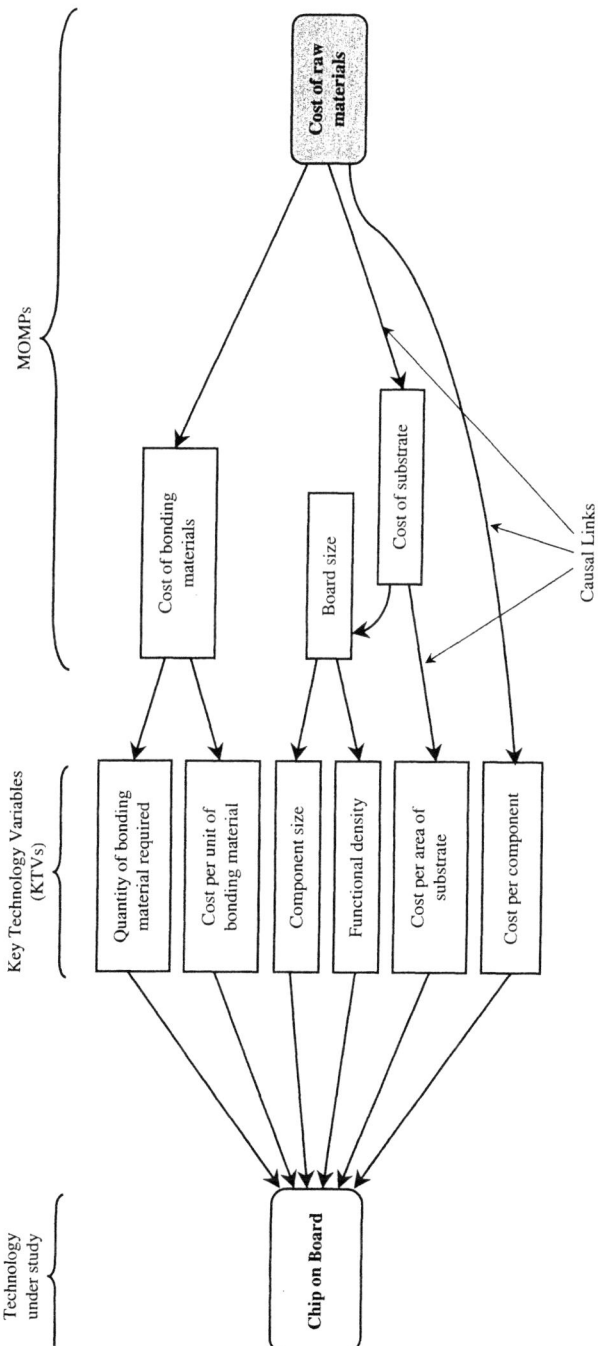

Figure 6.14 Fragment from the model showing causal links between COB and the 'cost of raw materials' MOMP

Analysing the sample Specific MOMP Map and the sample Specific Model

Once the specific MOMP map (Stage I of the AMBIT approach - chapter 4) and the specific model (Stage II of the AMBIT approach - chapter 5) have been constructed and validated, they can then be analysed. Their analysis, using the AHP approach is useful in distinguishing exactly what is and what is not important. The first phase of the analysis can consist of choosing the overall goal (i.e. Success of Company) and evaluating the options available. In doing so, the strategy manager can make all the necessary pairwise comparisons in order to determine the values for the relative priority of the specific MOMPs (i.e. specific MOMPs). Table 6.4 depicts sample priorities derived from Figure 6.5 shown earlier.

Table 6.4 Priorities of the specific MOMPs

Specific MOMP	Priority
WIP Cost	0.15
Cost of manufacturing process	0.45
Cost of raw materials	0.15
Quality of materials	0.125
Percentage late orders for substrate supplies	0.05
Percentage late orders for component supplies	0.075

According to Table 6.4 the Cost of Manufacturing Process is the key measure of manufacturing performance (MOMP) with respect to achieving the overall goal. If the user does not agree with the results, he can re-evaluate the specific MOMP map. Once the results are satisfactory, the analysis of the specific model follows.

During the construction of the specific model, when a causal relationship is created, it must be defined. For a constant causal relation this involves specifying the corresponding independent causal effect (ICE) and degree of confidence (DOC) values. In the case of a causal relation it involves specifying both the causal effect (CE) and confidence function (CF) functions. When the model has been completed and all the relations defined, the user can select the schema whose causal effects he wishes to propagate. In this example, by using the prototype tool (based on the theory described in Sections 6.2.1 and 6.2.2), chip on board (COB) can be selected as a leaf schema. Following this selection, the PROTIMA tool will then calculate the effect of COB introduction on the specific MOMPS. The sample output might look like the one depicted in Table 6.5.

In this example, the user is particularly interested in the Cost of Manufacturing Process, since it has the highest priority (Table 6.4). From Table 6.5, it is clear that the introduction of COB would result in an increase in the Cost of Manufacturing Process. The extent of the increase is 0.12 with a very high certainty that this indeed would happen (75%). However, as such an increase could have serious repercussions for an organisation that competes on cost, the user might decide to interrogate the system, in order to determine how it arrived at its conclusions. The fragment of the model connecting COB to Cost of Manufacturing Process is shown in Figure 6.15.

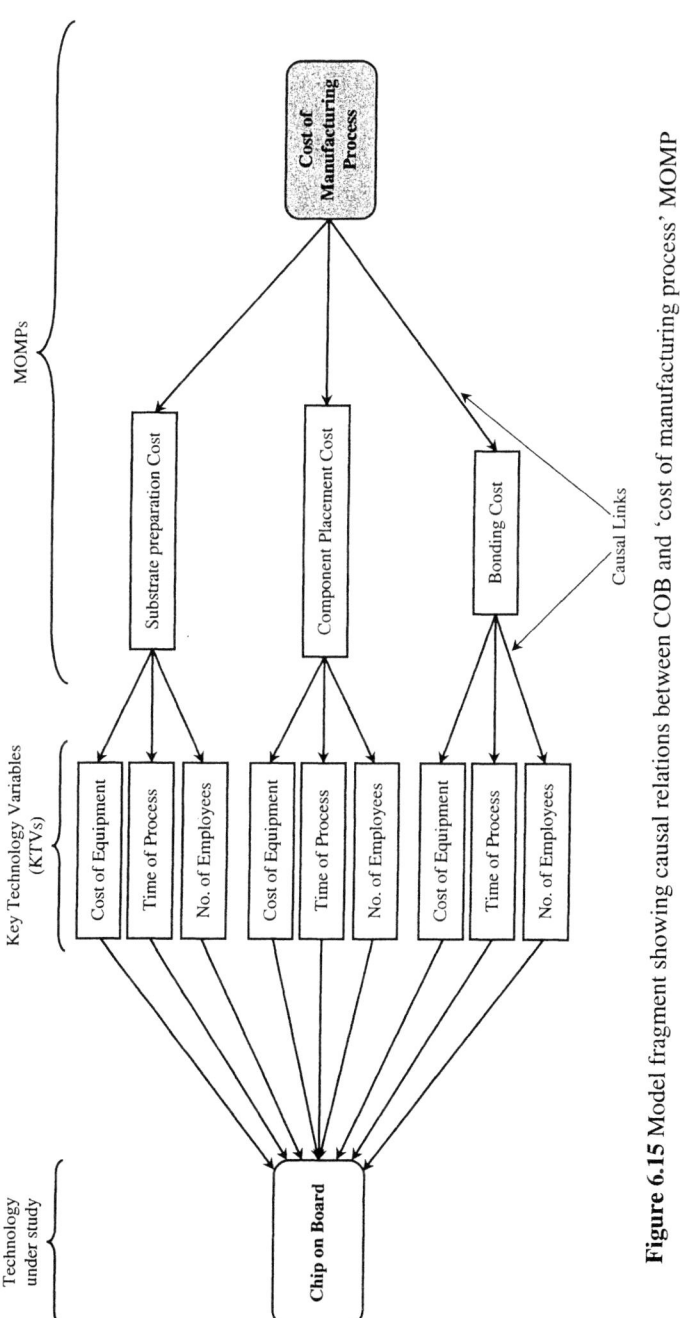

Figure 6.15 Model fragment showing causal relations between COB and 'cost of manufacturing process' MOMP

Figure 6.15 shows that the Cost of Manufacturing Process is causally affected by Substrate Preparation Cost, Component Placement Cost and Bonding Cost. In this figure, the manner in which the introduction of COB causally affects each of the nodes (i.e. Substrate Preparation Cost, Component Placement Cost or Bonding Cost) can be observed and further analysed.

Table 6.6 shows the extent of the effect and associated degree of confidence due to the introduction of COB on these nodes.

Table 6.5 Effect of COB introduction on the specific MOMPs

Specific MOMP	Effect	Extent of Effect	Degree of Confidence
WIP Cost	Decrease	0.02	0.73
Cost of manufacturing process	Increase	0.12	0.75
Cost of raw materials	Increase	0.16	0.90
Quality of materials	Decrease	0.19	0.66
Percentage late orders for substrate supplies	Increase	0.05	0.55
Percentage late orders for component supplies	Increase	0.04	0.53

Table 6.6 Linking MOMPs with their causal effects and associated degree of confidence

Linking MOMP	Effect	Extent of Effect	Degree of Confidence
Substrate Preparation Cost	Increase	0.20	0.65
Component Placement Cost	Increase	0.14	0.79
Bonding Cost	Increase	0.19	0.48

In tracing further back to find a reason for this overall cost increase in the cost of manufacturing process, it may be found that substrate preparation cost, component placement cost and bonding cost show a marked increase in Cost of Equipment. Meanwhile, they all show a slight decrease in Number of Employees and Time of Process. This post-analysis examination of the model may not cause the user to doubt any of his initial judgements, but may highlight the low degree of confidence in the predicted effect on Bonding Cost (i.e. 0.48, Table 6.6). The figures which have been used are the ones most likely to be the correct ones. They lie between what the user thinks are the best-possible and worst possible scenarios. Since the predicted change to Cost of Manufacturing Process is an unfavourable one, the user can decide to re-evaluate this portion of the model using the best-possible scenario figures, to see if they will alter the predicted change. As a result, the causal effects of COB on the key technology variables (KTVs) concerned with Bonding Cost (i.e. cost of equipment, time of process, number of employees) can be re-estimated.

The original Bonding Cost value (an increase of 19%) was estimated using the information as presented in Table 6.7.

Table 6.7 KTVs with their causal effects

KTVs	Effect	Extent of Effect
Cost of Equipment	Increase	0.7
Time of Process	Decrease	0.18
No. of Employees	Decrease	0.12

Similarly, the causal effect (CE) function for cost of equipment, time of process and number of employees can be estimated. A sample of the CE and CF functions for Cost of Equipment is depicted in Figure 6.16. The corresponding edited functions are shown in Figure 6.17.

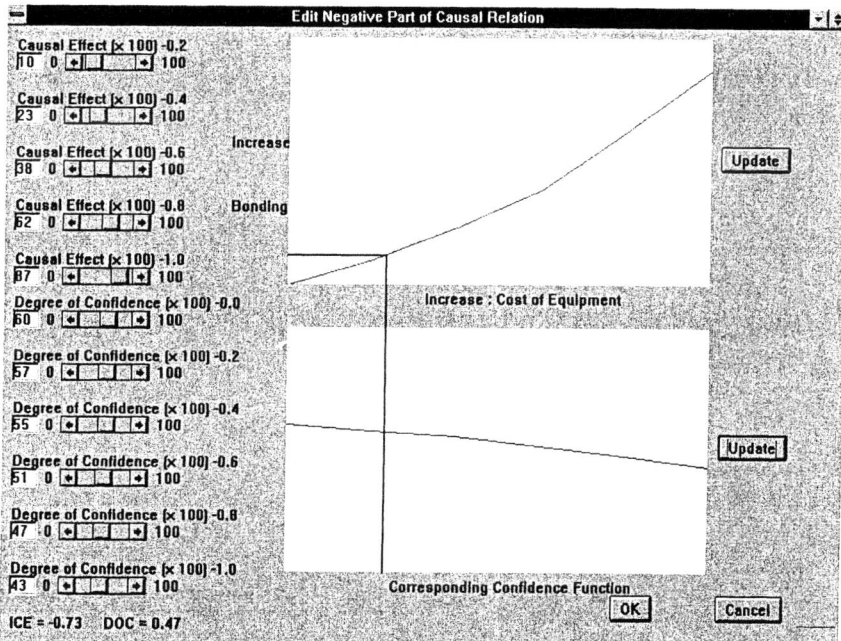

Figure 6.16 Initial CE and CF functions for cost of equipment

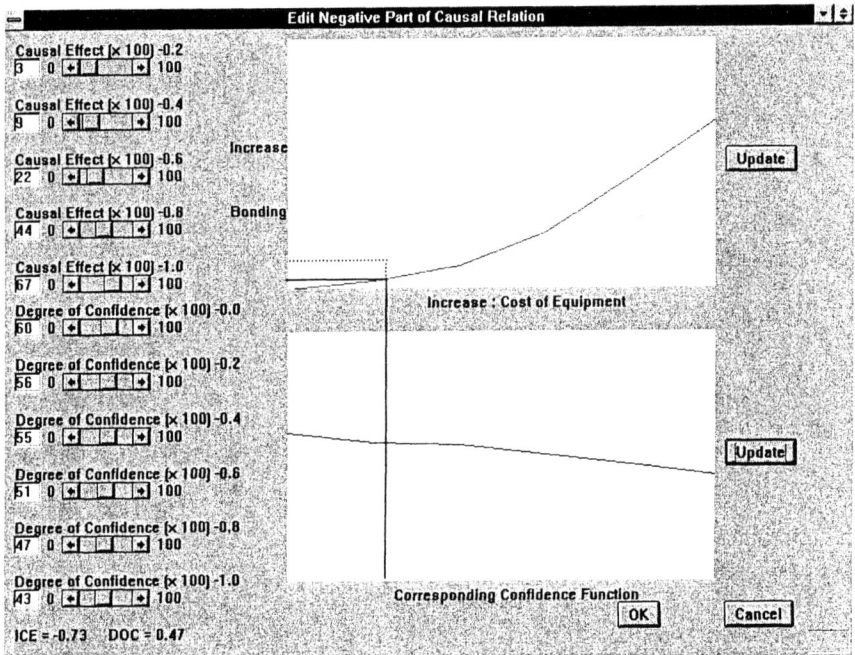

Figure 6.17 CE and CF functions for cost of equipment, after editing

In Figure 6.17, it is obvious that the new CE function is less steep and the corresponding confidence function is unchanged from the original. This means that the same increase in the Cost of Equipment KTV will result in a lesser increase in the Bonding Cost linking MOMP, than was the case with the old CE function (as shown by the dotted lines in Figure 6.17). With the new function, small increases in Cost of Equipment will not affect the overall bonding cost to a great degree. However, as Cost of Equipment increases further (as indicated by the slope of the curve), this KTV will have a serious impact on Bonding Cost. The causal effect functions for Time of Process and Number of Employees were similarly edited. Based on these edits, the overall effect to Bonding Cost is shown in Table 6.8.

Table 6.8 New figures for causal effects

KTV	Effect	Before CE edit		After CE edit	
		Extent of Effect	Effect on Bonding cost	Extent of Effect	Effect on Bonding cost
Cost of Equipment	↑	.7		0.6	
Time of Process	↓	.18	.19 ↑	0.25	.15 ↑
No. of Employees	↓	.12		0.15	

As shown in Table 6.8, an original 70% increase in cost of equipment, with an 18% decrease in time of process and a corresponding 12% decrease in no. of employees resulted in an overall 19% increase in bonding cost. However, after editing the CE functions, it was found that a 60% increase in cost of equipment, a 25% decrease in time of process and a corresponding 15% decrease in no. of employees now resulted in an overall 15% increase in bonding cost, which is more desirable than the original 19% increase.

This decrease in Bonding Cost, which is 4% (from 19% to 15%), would in turn propagate through to Cost of Manufacturing Process, resulting in a slight drop in the overall cost of manufacturing process. On examining the results, the strategy manager could conclude the following:

- Based on the organisation's mission statement, it is clear that the associated strategic performance indicators (SPIs) are cost in the manufacturing chain and quality in the supply chain.
- The decomposition of the SPIs results in the following specific MOMPs, WIP cost, cost of manufacturing process, cost of raw materials, quality of materials, percentage of late orders for components and percentage of late orders for substrate.
- When these MOMPs are prioritised (using the analytic hierarchy process (AHP) approach) in terms of their importance to the success of the company, the cost of manufacturing process is regarded to be the most important specific MOMP.
- The change from PTH to COB technology, would result in an increase in cost of manufacturing process, quality of materials and cost of raw materials.
- The strategy manager is fairly confident of the predicted increase in the cost of manufacturing process.

Taking all of the above implementation issues into account, it would appear that a move to COB technology would not be profitable to the organisation at this point in time. This is because such a move would adversely affect the organisation's key strategic targets i.e. keeping manufacturing cost to a minimum and ensuring that quality raw materials are delivered on time.

This example demonstrates how a tool such as PROTIMA could be used to examine the strategic impact of a new manufacturing technology or manufacturing programme on the business goals of an organisation. This example also shows how the PROTIMA based tool could allow the user to browse and interrogate the data in order to see how it reached its conclusions. The analysis of the specific MOMP map highlights the application of the analytic hierarchy process (AHP) technique, in determining exactly what is and what is not important. In addition to prioritising the MOMPs, AHP allows the user to browse the hierarchy, alter a single judgement and see the changes that propagate right through to the specific MOMPs as a result of a judgement alteration. Indeed, the method of pairwise comparisons is clearly an easier approach with which to work in comparison to the approach where the user is given 100 points and told to assign them to the MOMPs in proportion to their importance. How could such decisions be justified? Meanwhile, the analysis of the specific model shows how the KASPER tool propagates causal effects through a network, in order to help the user predict the effect that the introduction of COB would have on the specific MOMPs. The calculation of an associated degree of confidence for each causal effect is an additional intuitive feature, which can prove useful when there is some doubt or conflict concerning the data.

In synopsis, Section 6.2 shows how the AHP and KASPER analysis methods in the prototype PROTIMA tool, could support the user in identifying the strategic impact of introducing a new manufacturing technology or programme.

6.3   A Case Based Reasoning Tool

> CBR combines a cognitive model describing how people use and reason from past experience with a technology for finding and presenting such experience.
> – Choy *et al.* 2002

A second approach which is appropriate to AMBIT is case-based reasoning. Case Based Reasoning (CBR) supports the user in learning from the experience of earlier implementations of the selected technology or programme. When the user browses through existent cases and adapts the solution of these cases to the current problem, he learns from the experience of previous cases.

CBR represents a cognitive and computational approach of reasoning by analogy to past cases [Weber *et al.* 2003, Morris 1995]. Instead of using AI techniques which apply general knowledge of a problem domain or make associations along generalised relationships, CBR uses specific knowledge of previously experienced situations [Aamodt *et al.* 1994]. These situations are called cases. Each case generally contains a description of the problem, in addition to a description of the solution and/or the outcome. Other information includes a set of questions and answers that guide a user in identifying the closest matching case [Bagg 1997]. Thus, the term case describes a problem situation. A past case, previous case, stored case or retained case refers to a previously experienced situation which has been learned and stored, so that it can be used in the solution of future problems. Meanwhile, a new case describes a current problem to be solved. Cases are indexed so that they can be easily retrieved and then adapted to solve the current problem. The information and experience stored in a retrieved case is used to solve the current problem.

The basis principle of CBR is that decision makers often encounter problems which are not unique, but are instead variations of a previous problem. Clearly, it is easier for them to address the current problem by first using the solution of a previous related problem, than it is to generate the solution from scratch. It may happen that the decision maker will not find a case exactly matching the current problem. In this event the closest matching case, which can provide insight into how to solve the current problem, can be selected from the library of cases. One of the advantages of CBR in comparison to rule-based or model-based approaches, is that the concrete examples provided by the cases are much easier for users to understand and apply in various problem-solving contexts, than complex chains of reasoning generated by rules or models. Indeed, the CBR method of applying related historical knowledge to a current situation is an approach which humans often use to solve problems [Ross 1989]. "Case Based Reasoning is - in effect - a cyclic and integrated process of solving a problem, learning from this experience, solving a new problem etc." [Aamodt *et al.* 1994]. Thus, CBR focuses on incremental learning [Choy *et al.* 2002]. When a case is solved successfully, the approach is remembered and stored for application to future problems. However, when an approach fails, the reason for the failure is identified and remembered in order to avoid a similar mistake in the future. Consequently, the system learns as the user learns. For example, if a doctor's examination of a particular patient highlights similar symptoms of a previous patient, the doctor can use the diagnosis and treatment of the previous case and apply them to his current patient.

In a similar way, in the CBR approach, the current problem situation is first described. An historical case similar to this problem is then identified. This case is used to propose a solution to the current problem. Finally, the proposed solution is evaluated and the system is updated by learning from the experience (Figure 6.18).

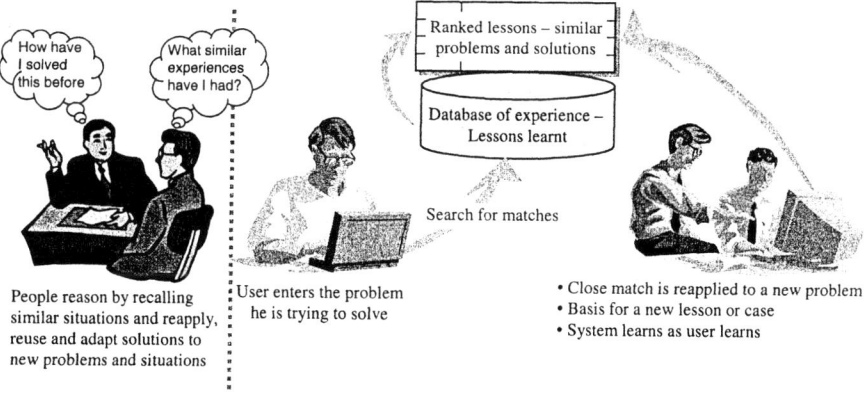

Figure 6.18 The Case Based Reasoning approach [after Bagg 1997]

At the highest level of generality, the processes in Figure 6.19 describe the CBR cycle. The new case, which is a description of the current problem, is used to retrieve similar historical cases from a library of previous cases. When a case is found which is similar to the current problem, it is called the retrieved case. Used in association with the new case and through reuse, it becomes a solved case.

The solved case basically represents a suggested solution to the initial problem. During the revise process, the solved case is tested, by applying it to the real world or having it evaluated by someone with the requisite knowledge. If necessary, modifications are made to the solved case. Whilst, the solved case is now referred to as the confirmed solution, the experience from this case is retained for future re-use, using the retain process. During retain, the case base is updated by the new learned case.

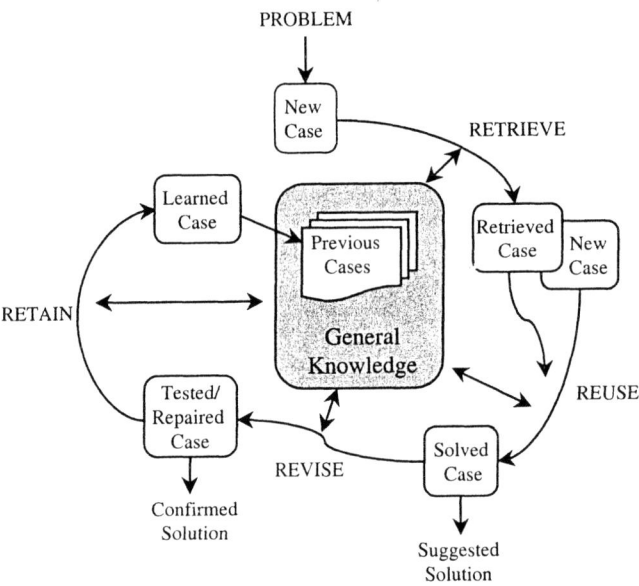

Figure 6.19 The CBR cycle [after Weber *et al.* 2003]

The principal activities in the CBR cycle, as depicted in Figure 6.19, can be described as follows:

- **Retrieve:** Retrieve the case which is most similar to the current problem. Retrieval starts with a (possibly partial) problem description and ends when a best matching case has been found.

- **Reuse:** Reuse the information and knowledge from the retrieved case and apply it to solve the current problem. Reuse focuses on identifying the differences between the retrieved case and the current case and identifying the part of a retrieved case which can be transferred to the new case. The solution of the retrieved case is generally transferred to the new case directly, as its solution case.

- **Revise:** Revise or adapt the case to solve the problem and then evaluate the proposed solution. The evaluation allows an opportunity to learn from failure.

- **Retain:** Retain the aspects of this experience which are likely to be of benefit in future problem solving and store them in the case library.

A more detailed description of the activities involved in the retrieve, reuse, revise and retain processes is presented in Figure 6.20.

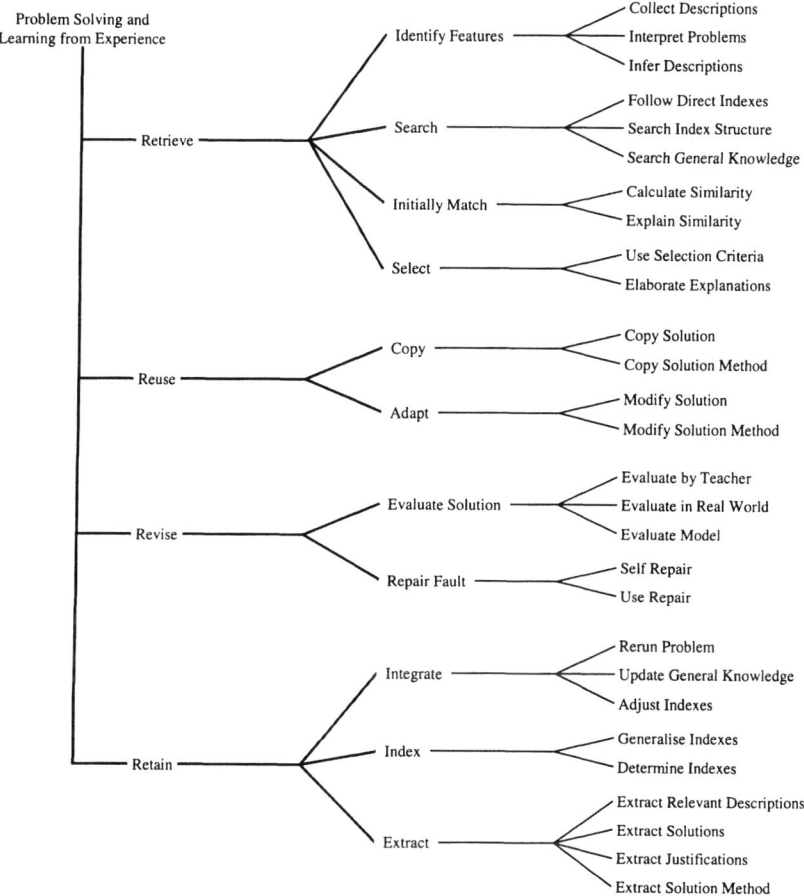

Figure 6.20 Activities of the main processes in CBR [after Aamodt *et al.* 1994]

Consequently, a CBR tool based on the AMBIT approach should include the key steps:

1. Retrieval,
2. Adaptation,
3. Validation, and,
4. Update.

In the remainder of this chapter, we will use sample screens from the laboratory prototype of the AMBIT CBR tool to illustrate the CPR approach as it might be used in a realistic software toolset. In this prototype, past cases are stored in a library of cases. These past cases record the situation before and after the technology or programme implementation, in addition to the actions necessary for successful implementation. Through the use of such historical cases, the user can gain insights into previous technology and programme implementations, in terms of the successes and failures of such implementations. Each technology and programme has a separate case base, which comprises several case files.

6.3.1 Retrieval

The process of retrieval in the prototype AMBIT CBR tool is shown in Figure 6.21.

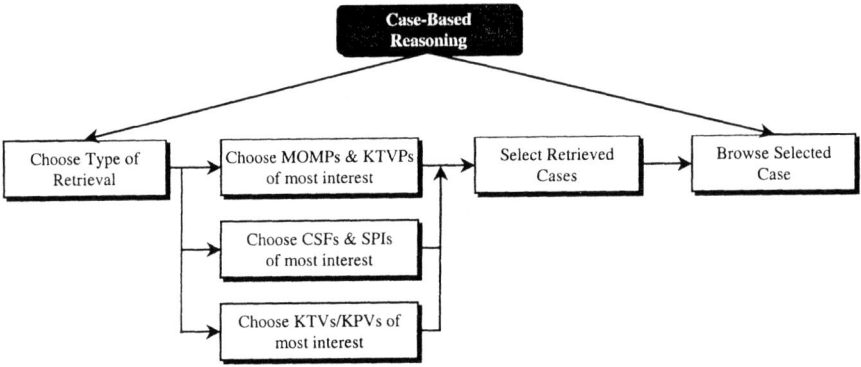

Figure 6.21 The process of retrieval

As is presented in Figure 6.22, the retrieval of the closest matching case to the current problem is based on; similar objectives, a similar situation or a similar implementation.

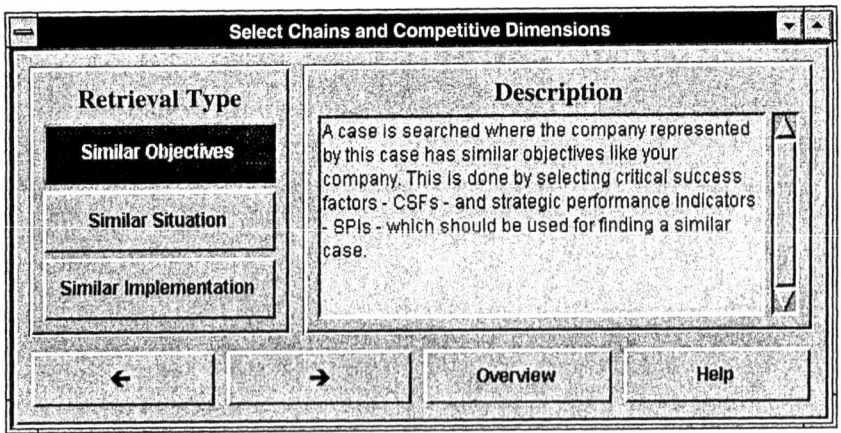

Figure 6.22 Select type of retrieval

In 'similar objectives retrieval', the objectives of the user's company should match those of a stored case in the case base. This means that as many as possible, of the stored case's critical success factors (CSFs) and/or strategic performance indicators (SPIs), should match the selected set of CSFs and SPIs of the company (described in the AMBIT framework). Figure 6.23 shows the list of CSFs and SPIs selected by the user during Stage I of the AMBIT approach. The user can select all or a subset of these CSFs and SPIs. On selection, a case having similar CSFs and SPIs (or as close a match as possible) is retrieved from the case base.

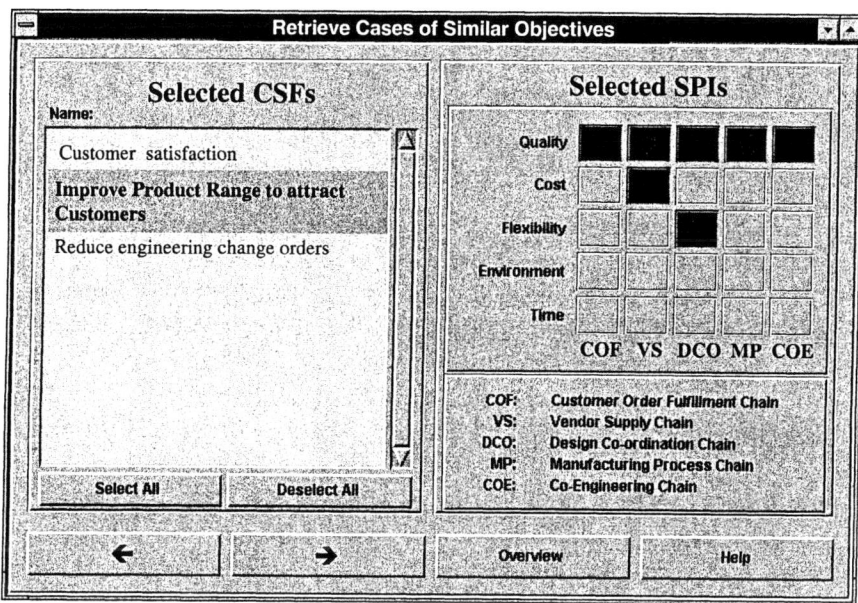

Figure 6.23 Retrieve cases of similar objectives

In 'similar situations retrieval' (Figure 6.24), the current state of the user's company should be similar to that of the company described in the stored case. In the context of the AMBIT framework, this means that the values of the selected MOMPs and/or key technology and/or key programme variables (KTVs/KPVs) should be similar to those of the retrieved case.

In this retrieval approach, a window displaying the names and current values of the MOMPs (selected by the user during Stage I) in addition to the names and current values of the KTVs/KPVs (selected by the user during Stage II) appear. If the user did not assign values to some of the KTVs/KPVs, he must do so, before the retrieval of case of a similar situation (in other words, the retrieval of a case having MOMP and KTV/KPV values similar to those selected by the user).

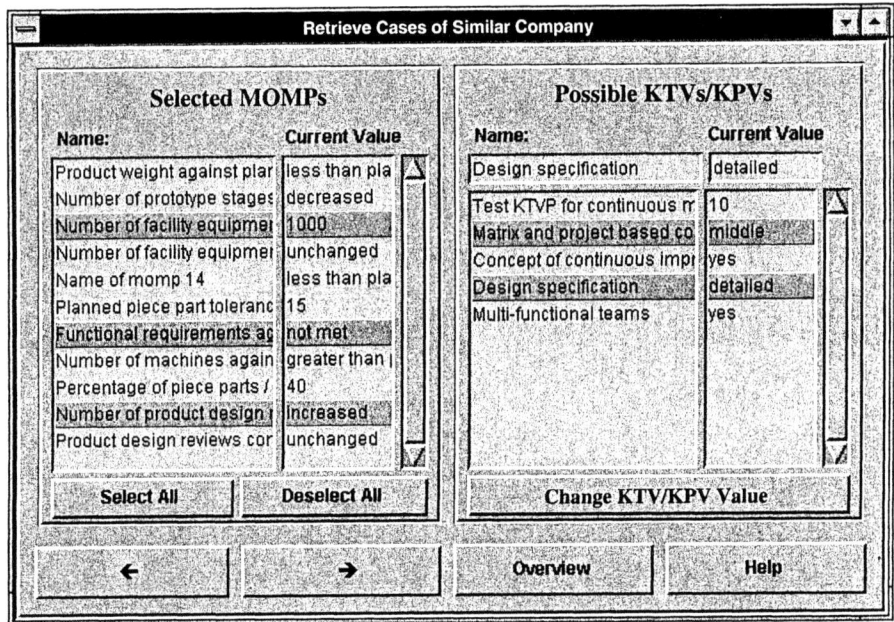

Figure 6.24 Retrieve case of a similar situation

In a 'similar implementation retrieval' (Figure 6.25), the user retrieves a stored case where a similar implementation has taken place. Once the user has extracted the stored case having the nearest matching implementation, he should be able to select all or a subset of this case's KTVs/KPVs and then experiment with changing their values and assessing the impact on the MOMPs and SPIs (Figure 6.25).

Once the user has identified the method of retrieval, the best-matching cases are retrieved and ranked. In the retrieval of cases with similar objectives, rank is calculated by counting equal CSFs, SPIs and MOMPs. Once the ranking value has been assigned to the retrieved cases, the case with the highest ranking value is generally selected, as this will represent the closest-matching case.

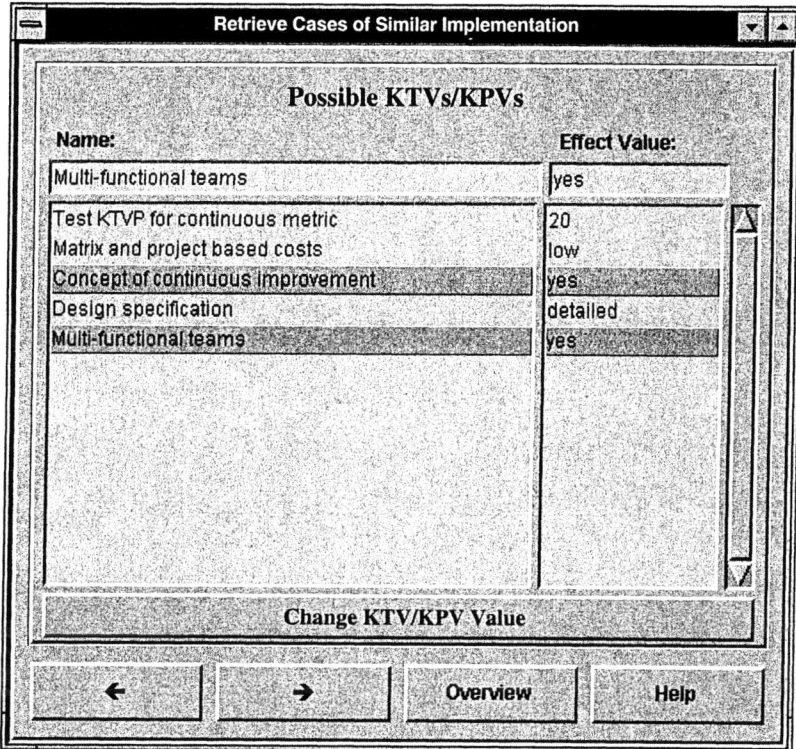

Figure 6.25 Retrieve cases of similar implementation

Once the user has identified the method of retrieval, the best-matching cases are retrieved and ranked. In the retrieval of cases with similar objectives, rank is calculated by counting equal CSFs, SPIs and MOMPs. Once the ranking value has been assigned to the retrieved cases, the case with the highest ranking value is generally selected, as this will represent the closest-matching case.

6.3.2 Adaptation

Kolodner suggests that there are two styles of case-based reasoning [Kolodner 1993]:
1. Cases which provide suggestions of solutions to problems i.e. problem-solving case-based reasoning.
2. Cases which provide context for understanding or assessing a situation i.e. interpretative case-based reasoning.

Despite a certain amount of similarity, no case is ever exactly the same as the problem of the user. Understandably, no user can expect his organisation to be affected in exactly the same way, even if he takes actions similar to those portrayed in the historical case. Clearly, no organisation, no matter how similar its goals, is a mirror image of any other organisation. Because of this difference, even despite a certain amount of similarity, the best matching retrieved case will more than likely require adaptation.

During adaptation, the user browses the case, in order to more fully recognise and understand the relationships between the objects of the problem domain (Figure 6.26). Because browsing and therefore adaptation increases learning, the user is, as a result, more capable of predicting the influence of a technology or programme implementation using the known effects which appeared in the retrieved case. Prediction is supported only if the retrieved case is totally similar to the current situation.

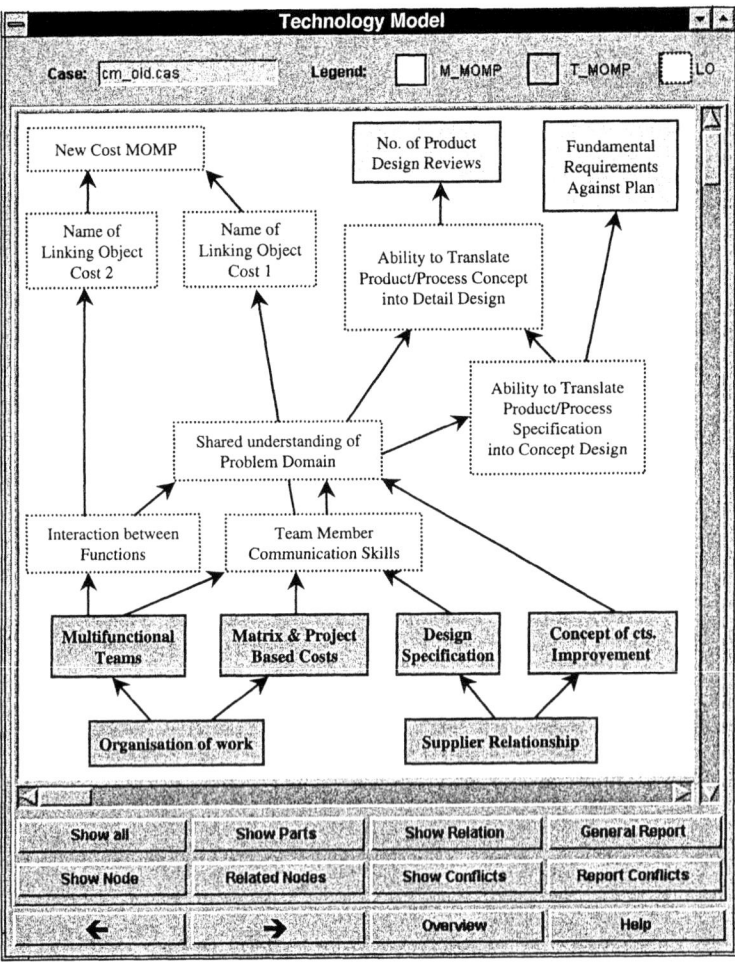

Figure 6.26 The technology model - browsing the model

The technology model window (of Figure 6.26) displays the network which the user can browse in order to support the process of adaptation. This network consists of nodes (e.g. MOMPs, KTVs and KPVs) and their respective relations.

6.3.3 Validation

Validation requires that the solution is validated either through feedback from an expert or from its application to the real world.

6.3.4 Update

The case base is updated with the solved case.

6.4   Conclusion

Stage III of the AMBIT approach is concerned with developing analysis tools to support the strategy manager in evaluating the strategic effect of a particular technology and/or programme implementation and/or change.

The prototype PROTIMA (**Pro**gramme/Technology **Im**plementation **A**nalyser) tool incorporates two methodologies: KASPER (Knowledge Based Advisor and Strategic Programme Evaluator) and AHP (The Analytic Hierarchy Process). KASPER allows the user to assess the impact of changes to the technology and programme on the specific MOMP map. It does so, by supporting the user in qualitatively and qualitatively reasoning across the specific model. AHP is used to prioritise the elements within the specific MOMP map hierarchy. It therefore reasons across the specific MOMP map.

The prototype AMBIT CBR (Case Based Reasoning) tool uses a cognitive approach of reasoning by analogy to past cases. In the CBR tool, a previously experienced situation which has been learned and stored, and which is similar to the current problem is retrieved, so that it can be used in the solution of the current problem. The user is then encouraged to browse the information and experience stored in this retrieved case. Browsing supports the user in learning from the experience of the old case. It also provides insight into ways in which to solve the current problem.

In developing a software tool based on the AMBIT approach, the authors recommend that a reporting module be a core element of Stage III. The reason for this, is that the design and structure of this module will guide the user in navigating through the AMBIT methodology and facilitate him in adopting the best course of action. In any software-based tool that requires active human interaction, the design of the user interface is critical. The user interface should not only *guide* the user through the various steps of the approach but also provide him with a greater insight and understanding of the complexities and intricacies of the underlying knowledge. Additionally, the graphical user interface should be able to accommodate the learning profile of the user and promote the effective dialogue between him and the system. Chapter 7 tackles some of the issues involved in developing graphical user interfaces.

Chapter 7

Learning and Creativity

Keywords: GUI, cognitive models, learning, CBLE and creativity.

Chapter Objectives: Management tools based on the AMBIT approach should be learning tools. In the design of an AMBIT software toolset, it is important to ensure that management learning is facilitated through the use of the tool. Thus, in this chapter we look at the theory of learning and that of creativity, and their impact on user interaction with IT tools. These areas need to be examined in order to develop an approach to the design of a Graphical User Interface (GUI), which is an integral part of any managerial learning tool. Having read this chapter, the reader should become familiar with:

- The importance of the graphical user interface (GUI).
- The Kolbian learning model and the learning style personalities arising from the four approaches to learning.
- The need to design the AMBIT approach based tool such that it operates within the context of a computer based learning environment (CBLE).
- The importance of creativity and the need to cater to the enhancement of creativity in GUI design.

7.1 Introduction

> In Britain, manufacturing firms that recognise the need to be at the forefront of creativity and technology, are proving well placed to take advantage of the opportunities afforded by the global marketplace. — DTI 2002

Whilst instantaneous global information is accessible at our fingertips, computers allow us to store and access this information relatively easily. They support communication with local, organisation wide and global networks. With the right graphics packages, we are able to manipulate and visualise n-dimensional data in a variety of different formats. Computers can influence the basis of the thinking process by extending our memory and our ability to work with data, information and ideas. However, given the possibilities of computers, and the investment of software houses in IT tool design and development, the power of this information explosion relies predominantly on the user's ability to absorb and react to the data and information. It is increasingly apparent that the major obstacle to scientific and economic growth may reside within humanity's limited ability to absorb and apply new information. Thus, given that computers can make information accessible and to a large extent the usability of an information system is dependent on the effectiveness of the user interface, society's ability to absorb and use information is affected by the graphical user interface (GUI).

A GUI gives incredible potential for enhancing the manner in which users interact with computers. Whilst the GUI is increasingly regarded as the main gating function to the popular use of information technology to a wider audience, Constantine points out; that "some of our efforts to improve may only have made matters worse. We have been rushing to deliver the latest 3D look-and-feel, only to have ungrateful users complain they can't read our grey-on-grey dialogues" [Constantine 1998]. Therefore in this chapter, we will discuss the following topics; learning styles, creativity ("today's tools often contain interface elements that stymie creative efforts" [Selker *et al.* 2002]), computer based learning environments (CBLEs) and AMBIT within the context of a CBLE.

7.2   Graphical User Interfaces (GUIs)

> Despite many technological advances that have facilitated better interface design, the daily experience of using computers is still fraught with difficulty and barriers for most people.   – Batra *et al.* 1998

An interface basically describes the place where contact between two entities occurs (e.g. a keyboard is an interface between a user and a computer). The more dissimilar the entities, the greater the need for an effective interface. The usual purpose of the GUI was and is to direct, orchestrate and focus the user on the tasks at hand through seamless interaction. Whilst the GUI includes every interaction that a user has with a computer, an effective GUI is one that the user continues to use and which has the following attributes [Roy *et al.* 2001]:

- Ease of navigation.
- Consistency.
- Learnability.
- User guidance or support.

As computer technology has become more widely available, the need for accessibility and ease of use i.e. the need for effective GUIs, has become increasingly pronounced. Today's users have become discriminating and intolerant of badly designed and ineffective GUIs. Their impatience with unfriendly systems and their expectations of more usable technology means that they are no longer willing to adapt to the system. They demand that the system adapt to them. What is thus required is a GUI which conforms to the user's behaviour and cognitive model of the system. GUIs undoubtedly give incredible potential for enhancing how users operate with computers. However, rather than minimising interface concerns, interaction standards have been developed to actually increase the amount of mental work it takes to use a GUI. As Rosenstein claims "in software and web design, we've all seen examples of user interfaces that are supposed to make our work easier, but which are actually so frustrating to use that we avoid them whenever possible" [Rosenstein 2001]. Whilst organisations have demonstrated their ability to produce software products with complex functionality, they have not equally demonstrated their ability to deliver more value [Constantine 1998]. The fact that usability is receiving increasing attention is evidenced by the proliferating number of books on user interface design. "All the major software houses have elaborate user interface testing programmes supported by expensive usability labs. We lavish more and more attention on user interface programming – as much as two thirds or some project budgets – and still miss the mark by miles" [Constantine 1998].

Graphical user interface (GUI) design is a critical skill and good GUI designs require that the developer learn and apply some basic principles, including making the design something with which the user will enjoy working. Good GUI design does not just imply 3D graphics, amazing splash screens, great animations and floating toolbars. It is about supporting the user in what he wants to accomplish in as easy and efficient a manner as possible [Nielsen 2001]. Once the designer has analysed and validated the informational requirements of the targeted user/users, design becomes the core element of usability. However, an important consideration in design should be the fact that most people have strengths and weaknesses in learning. Each of us learns in different ways [Bal *et al.* 2001]. If someone is building and designing tools to assist us in the way in which we process information, it is important that learning differences are taken into account, as a learning style refers to the way in which individuals acquire and use information [Karuppan 2001].

7.3   Cognitive Models

> An individual's cognitive style is a characteristic property of his/her own attitude or preference for information. It is consistent over time and is not influenced by situational factors within a normal range of conditions.   – So *et al.* 2003

Although there has recently been an intensive growth in technology, Buxton claims that human capabilities have not progressed at a similar rate [Buxton 1995]. Thus, whilst systems are becoming increasingly complex, the people using them still have the same abilities and limitations. In the past, this problem was generally resolved by simplifying the GUI design. Today, the focus is concerned with accelerating the transition from a novice to an expert user, a transition which is supported through the close correlation of the user's conceptual system model with actual system operation.

Schactman defines a model as "a generalised, hypothetical description used in analysing or explaining something; a simplified representation of a system, intended to enhance our ability to understand, predict, and control the behaviour of the system" [Schactman 1996]. However, although designers may carefully develop a conceptual system model which they believe corresponds to the user's system vision, users can often approach such systems with several inconsistent, diversely sourced mental models. Furthermore, the user of an unfamiliar system might turn to these accustomed models when he encounters uncharted functionality, with the outcome frequently resulting in confusion and errors. Conversely, if the user has built up a good cognitive model of the system, user interaction will be facilitated.

Cognitive models or cognitive maps are terms used to describe conceptual representations that are created as a means of explaining the operation of complex systems. Whilst the cognitive model represents how the person believes the system works, the more a user comprehends the operation of the software, the more accurate the cognitive model he creates and the more quickly he learns the effective and efficient use of the system (Figure 7.1).

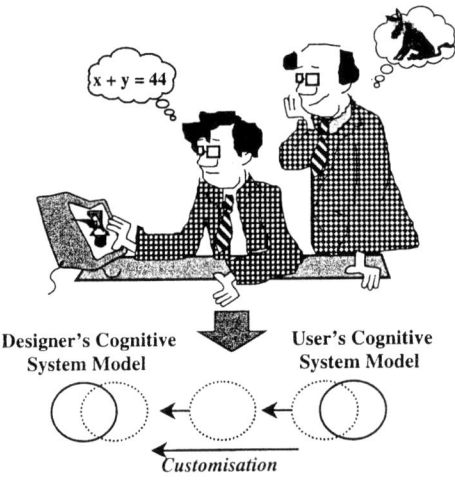

Designer's Cognitive   User's Cognitive
System Model   System Model

Customisation

Figure 7.1 Cognitive system model mapping with respect to system functionality

If the users are supported in customising each application they use, in order to reflect their cognitive system model, inconsistency across different tools is as a result reduced. Basically, users who are involved in the design or customisation of the system GUI are more familiar with the operation of the system and should then feel in more direct control of the system interface.

Notwithstanding, users should also be supported in making "their own task the exclusive locus of their attention, by designing the interface such that it can be reduced to habitual operation" [Raskin 1997]. This approach applies to the field of experimental psychology, which suggests that we can pay attention to only one thing at a time. Thus, if we are involved in doing two or more activities concurrently, then all of these activities bar one should be habitual. However, the problem with current user interface designs is that they demand "that we must be consciously aware of both the task at hand and the current system state" [Raskin 1997].

The increased user-focus emphasis in the GUI domain reflects developments in the software design domain. These developments are emphasising the importance of the GUI discipline. Indeed, the current emphasis of software interface design on people, their circumstances, what they require and how they interact with the software, is ensuring that "the richness of human experience comes to the foreground and computing sits in the background in the service of these experiences" [Winograd 1995].

7.4 Theory of Learning

> People vary in how they take information in and how they transform the information into meaning. — Buch *et al.* 2002

In catering to global information access, IT tools and technologies to augment learning and human memory are required. For information to be assimilated it must be moved from working memory into long-term memory. Whilst long-term memory has a large capacity, working memory, is characterised by its limited capacity. As a result, working memory represents the bottleneck in human information processing. Thus, information in working memory must be used else it becomes lost. Within the context of cognitive psychology, the technical term used for this process is Rehearsal. Rehearsal occurs when a user forms a visual image or re-organises information for the purposes of future reference. It is effective once information has been encoded in long-term memory [Clark 1995]. We believe that individual user participation in system design and system customisation supports information transferral from working to long-term memory. Furthermore, cognitive psychology offers the following support for this premise; when a user is learning a new interactive system, the whole process may be aided or impeded by what he may already know from other systems.

Karuppan claims that a correlation exists between an individual's learning style and his strategy in processing information [Karuppan 2001]. Thus, through the use of tools which are designed to accommodate his preferred learning style and customised to reflect his cognitive model, one could surmise that the learning curve will be reduced and interaction facilitated. Direct user involvement can induce the *Hawthorne Effect*, which suggests that if people sense that they are involved in something over which they have some control then they tend to work harder and learn faster.

With reference to learning theory, the authors will primarily draw on the work of David Kolb[14] who claims that the need to adapt is the driving force behind learning. Hence learning, which occurs at all life stages, may be described as the process whereby knowledge is transformed by experience [Kolb 1984, Lefrancois 2000]. Kolb, on defining two sets of opposed orientations within the learning cycle, identifies learning as the result of conflicts arising from these opposed ways of dealing with the world. Thus, the manner in which the conflict is resolved between these orientations determines the resultant level of learning. The opposed orientations are: abstract conceptualisation versus concrete experience and reflective observation versus active experimentation. Abstract conceptualisation and concrete experience are associated with comprehension and apprehension respectively i.e. they represent the manner in which the experience is acquired. Comprehension refers to the reliance on conceptual and symbolic representation, whilst apprehension is more associated with that which is tangible. Once, the experience is grasped, it is then transformed through either intention (reflective observation) or extension (active experimentation). The intentional transformation essentially describes the internal reflection of the experience. Meanwhile, extensional transformation refers to the active and external manipulation of the outside world. Experience grasping and its subsequent transformation results in knowledge creation [Kolb 1984].

Kolb asserts that if the learning process is to be efficient then the learner requires four different types of learning skills (namely: experience, reflection, abstract conceptualisation and experimentation), to be adopted alternately to the learning problem. Hence, the ideal learning process includes moving continually from concrete experience (becoming involved in new experiences), to reflective observation (reflecting on these experiences), to abstract conceptualisation (integrating these observations into models and theories), to active experimentation (using these theories in the resolution of problems) and then back again to concrete experience [Isaacs *et al.* 1992]. The combined use of these four styles of learning represents the highest form of learning. However, this scenario does not always occur, for in general people have one dominant learning style [Rollins 1993].

The learning style personalities arising from the opposed orientations are as follows [Kolb 1984, Buch *et al.* 2002, Honey *et al.* 1992, Frontczak *et al.* 1991] (Figure 7.2):

1. **Accommodators** are generally open-minded, patient and tolerant. Whilst, they enjoy doing things and participating in new experiences, they tend to solve problems in an intuitive trial-and-error manner, relying heavily on other people for information rather than on their own analytic ability. Accommodators focus on feeling and doing. Given the characteristics of accommodators, the following are applicable to this group:
   - Hands-on computer-based simulation games [O'Conner 1998].

---

[14] Kolb's learning styles inventory is one of the most widely used instruments in identifying students' chosen strategies to process information in learning and problem solving situations [Karuppan 2001].

- Self-directed learning methods such as role plays, games, observations, and simulations [McCarthy 1980].
- Computer-based training [Birkey *et al.* 1995].

2. **Divergers** excel at viewing concrete situations from many perspectives and organising many relationships into a meaningful 'gestalt'. The greatest strength of the diverger lies in his imaginative ability, information gathering skill, and his awareness of meanings and values and the hidden implications of ambiguous situations. Divergers emphasise feeling and watching. Given the characteristics of divergers, the following are applicable to this group:
   - Brainstorming [Motter-Hodgson 1998, Blackmore 1996, Dixon *et al.* 1995].
   - Tasks and activities involving reflection [Motter-Hodgson 1998].
   - Sufficient feedback [Motter-Hodgson 1998].

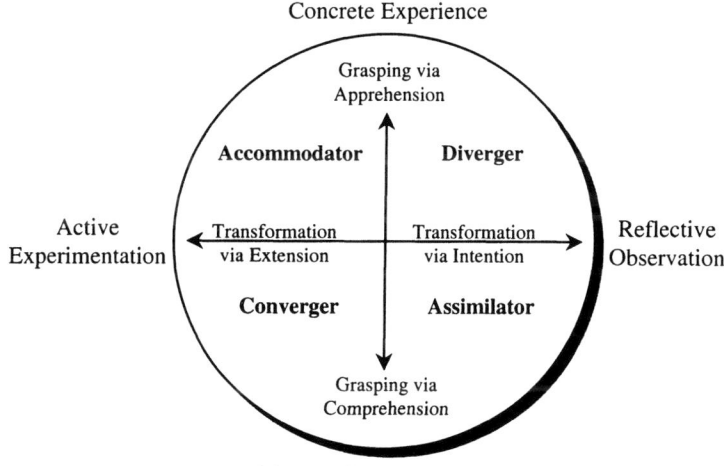

Figure 7.2 The Kolbian learning circle [Kolb 1984]

3. **Assimilators** favour ideas, intuition, seeing relationships between things and inter-related meanings. They excel at organising information, building conceptual models, testing theories and ideas, designing experiments and analysing quantitative data. Assimilators focus on thinking and watching. Given the characteristics of assimilators, the following are applicable to this group:
   - Group exercises and simulations impede rather than enhance their learning experience [O'Conner 1998; Richter, 1992].
   - Information should be presented in a systematic and organized sequence [Blackmoore 1996; Buch *et al.* 2001].

4. **Convergers** can generally be characterised by their logical, analytical, objectively critical and efficient approach to problems. Their greatest strength lies in problem solving, decision making and the practical application of ideas.

The Converger attitude is based on thinking and doing. Given the characteristics of convergers, the following are applicable to this group:
- Small-group discussions [Richter 1992].
- Ability to judge their own work and have that work evaluated as simply right or wrong [Krahe 1993].
- Ability to generate singular solutions to problems [Krahe 1993].
- A learning environment which supports trial-and-error experimentation [Felder 1996].
- Data-based programmes which offer the capacity to conduct tests that immediately illustrate various scenarios [O'Conner, 1998].

When an individual has a pronounced learning strategy, he tends to combine adopted subsidiary learning styles with the preferred learning approach. These subsidiary styles are embraced at various behavioural levels from personality to task related skills. Indeed, the career an individual pursues may expose him to a learning environment and manner of thinking which is necessary to success within that career niche. Working with a group of people who share a common mentality and set of beliefs will generally involve the incorporation of those beliefs into the learning profile [Morgan *et al.* 1992, Kolb 1984]. Also, the demands and pressures of a task tend to influence the individual's learning orientation.

Kolb categorised the individual's functional role within the organisation as follows [Kolb 1984]:
- The Executive role requires a strong orientation towards decision making and task accomplishment within an uncertain environment e.g. general management. This task role is best suited to the Accommodator.
- The Personal role characterised by inter-personal communication and inter-personal relationships is best assumed by the Diverger.
- The Informational role represents data collation and analysis e.g. planning and research, with the Assimilator being most suited to this task.
- The Technical role which requires both technical and problem solving skills is best undertaken by a Converger.

The ideal learning environment is thus one in which the learner cycles through the four stages as identified by Kolb i.e. having an experience, reflecting on this experience, integrating these reflections into models and theories and then using these theories in the resolution of problems. However, as already pointed out, it is generally the case that a person will use one learning style exclusively to the detriment of the other learning styles.

The authors realise that whilst it is necessary to cater to a person's predominant Kolbian learning style, whilst encouraging the lesser-used learning styles in order to foster a more 'optimum' learning experience, it is necessary to view the macro learning frameworks in which the kolbian learning styles reside. The authors believe that these macro learning frameworks are as follows; ([Beaver 1998, Bal *et al.* 2001] – modified).

- *Auditory learning framework.* Auditory learners are also known as verbal learners. They learn more from written and spoken explanations.
- *Visual learning framework.* Visual learners absorb information more effectively from pictures, diagrams, films and demonstrations.
- *Kinaesthetic learning framework.* Kinaesthetic learners prefer learning through their bodies. They excel in viewing relationships. The whole area of virtual reality (VR) and immersion in a virtual world would be ideal for the kinaesthetic learner.

It is important that the macro learning framework for an individual is recognised. As an example, if a thing graphs, models, diagrams are explained verbally to a visual individual, then his learning is not 'optimised' as he will unconsciously have to translate those words into mental imagery.

Combining the macro learning framework with the kolbian learning styles, one could have the situation whereby a person may have divergent characteristics within a visual learning framework. As an example, this person may excel at viewing situations from many perspectives but requires that the presented information be predominantly graphical in order to support the identification of meaning and inter-relationships and ultimately support his learning preference.

Notwithstanding learning styles, the design of systems should consider the fact that the learning process is supported when [Beaver 1998]:
- We are in safe supportive surroundings.
- All our neurology is engaged in the process.
- We are allowed to learn in our own way.
- Our conscious mind is distracted.

> To ignore differences in the individuals hinders the conclusions of any research that considers the impact of information systems.   – So *et al.* 2003

7.5   A Computer Based Learning Environment (CBLE)

A computer based learning environment (CBLE) provides decision makers with greater scope of learning through conceptualisation, experimentation and reflection. Whilst tools of a CBLE nature allow decision makers to avail of opportunities which are not easily achievable in daily management activities, they also assist in "providing a structured way of thinking about complex problems" [Isaacs *et al.* 1992]. The increasingly mutable nature of the organisational environment, whilst commonly prohibiting the testing of theories and the evaluation of conflicting ideas, generally requires that managers adopt a 'hands on' learning attitude. The learning cycle as it operates in ideal management situations, comprises:
- Discovery of problems (mental models).
- Invention of solutions (strategy and decision making).
- Solution implementation (real world implementation and action).
- Reflection on the impact of solution implementation (outcome and evaluation) (Figure 7.3).

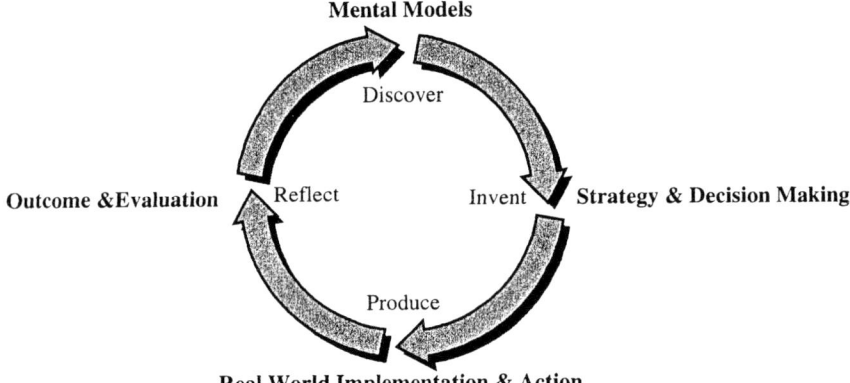

Figure 7.3 Ideal management learning cycle [Isaacs *et al.* 1992]

In an ideal environment, "new insights about the real world would be continually discovered and embedded in improved mental models, decisions would be invented based on new mental models, those decisions would be enacted and then outcomes would be reflected upon to produce new insights" [Isaacs *et al.* 1992]. However, cognitive and perceptual sets, in the form of implicit mental models limit mental model improvement. Delays between strategic decision making, implementation and evaluation may be protracted. Differences in the mental models of decision makers can result in different interpretations of similar data. Unanticipated changes in technologies, markets and/or economic conditions affect the impact of strategic decisions, especially if made over a long time span. Such factors constrain managerial learning. However, a CBLE addresses these issues, for within such an environment, time can be slowed down or accelerated, irreversible actions can be made reversible and the risks which are normally associated with experimentation in the real world, can be eliminated. A CBLE provides a virtual world (Figure 7.4), where actions can be implemented instantaneously and their outcomes assessed rapidly.

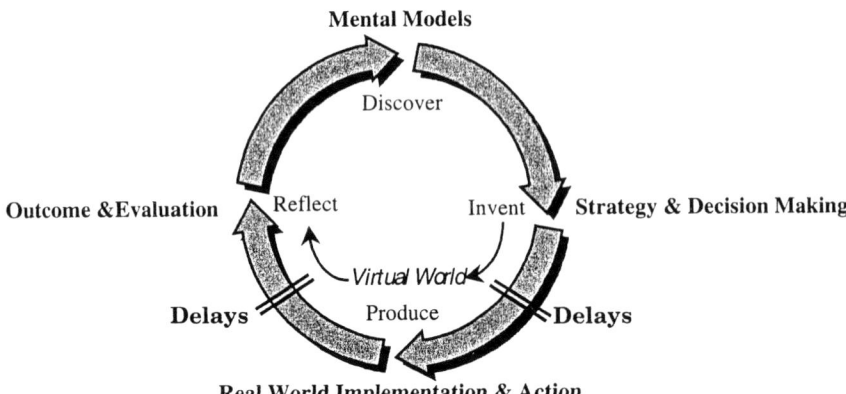

Figure 7.4 A CBLE learning cycle [Isaacs *et al.* 1992]

7.5.1 AMBIT as a CBLE

As a result of increased competition and the need to constantly keep abreast of changes, management's ability to pause for substantial reflection is seriously restricted. We believe that the AMBIT approach, within the context of a CBLE, fosters management learning and reflection, in the following ways:

- **A CBLE can expose gaps between the way management believes the organisation works and the way it actually works** [Isaacs *et al.* 1992]. In the AMBIT approach, manufacturing strategy is expressed in terms of the critical success factors (CSFs) of the company (e.g. increase product range). As already outlined strategic performance indicators (SPIs) are measures of performance which are used to measure the individual business processes in terms which are appropriate to/compatible with the organisation's CSFs. SPIs relate to the business processes (manufacturing, co-engineering, design co-ordination, vendor supply chain and customer order fulfilment), the macro measures of competitive performance (time, cost, quality, flexibility and the environment) and the manufacturing typology (make to stock, assemble to order, make to order and engineer to Order). Strategic performance indicator (SPI) identification is really the identification of the business process and associated competitive dimensions, which have to be considered in realising the critical success factors. Thus, the identification of the organisations' strategic performance indicators (SPIs) (Figure 7.5) serves to enhance managerial awareness and understanding of the organisational goals and organisational framework.

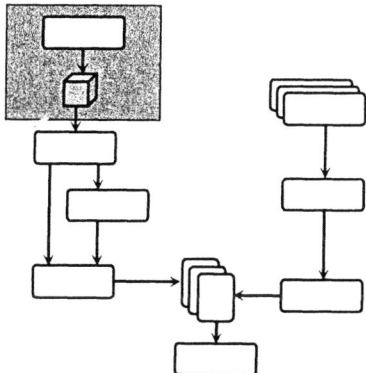

Figure 7.5 A CBLE can clarify management thinking

- **A CBLE provides the decision maker with new opportunities for learning through conceptualisation, experimentation and reflection.** An AMBIT based tool should enable management to observe how their current technologies (e.g. Interconnect technology) and programmes (e.g. Concurrent Engineering) and alternative technologies and programmes impact the organisational goals. Learning opportunities are provided by supporting:

○ MOMP map browsing and customisation. MOMPs (see Chapter 5) as a result of their linkage to SPIs, represent measures of manufacturing goals, coherent with the overall business goals of the business strategy.
○ Technology map browsing and customisation.
○ Analysis of the impact of current technologies and programmes or their alternatives, on the organisation's goals and overall business strategy (Figure 7.6).

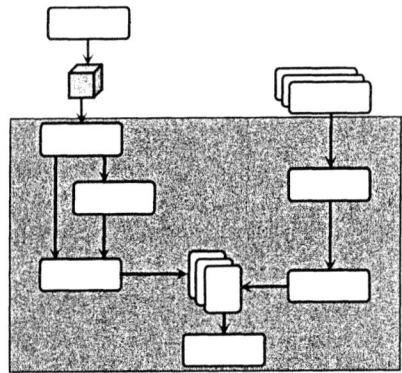

Figure 7.6 A CBLE provides new opportunities for learning

By allowing decision makers to assess the outcome of technology and programme adoption, decision makers are supported in understanding the complex interactions that can produce persistent problems.

Experimentation encourages awareness of the manner in which different strategies are needed to achieve desirable behaviour in the organisation as a whole. Xiao *et al.* maintain that fixation errors often result in decision making within dynamic environments (i.e. the decision maker usually has to make quick and snap decisions). Fixation errors occur if the decision-maker fixates on cues and hypotheses incorrectly, or if he is obsessed with a course of action and then becomes indisposed to changing this course despite conflicting evidence [Xiao *et al.* 1995]. Evidently, alternative scenario and hypotheses experimentation in a risk-free environment mitigates the occurrence of fixation errors.

■ **A CBLE supports the development of new mindsets into the nature of the organisation.** Experimentation and browsing endorse alternative scenario investigation and mental model exploration (Figure 7.7). Rapid feedback from experimentation facilitates insight into the consequences of managerial decisions and supports the development of new mind-sets.

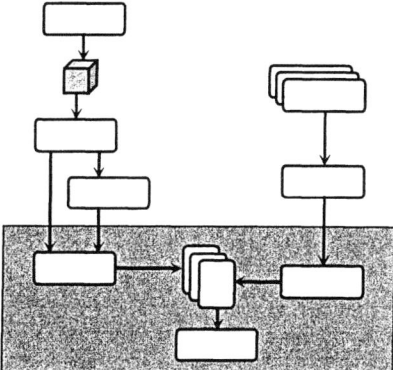

Figure 7.7 A CBLE supports the development of new mindsets

- **A CBLE supplies an integrative demonstration of the complex and subtle dynamics at the micro and macro organisational levels.** Use of an AMBIT based tool should enable the user to view the impact of technology and programme adoption at the micro level (measures of manufacturing performance – MOMPs) to the macro level (strategic performance indicators – SPIs) and subsequently the goals of the organisation (Figure 7.8).

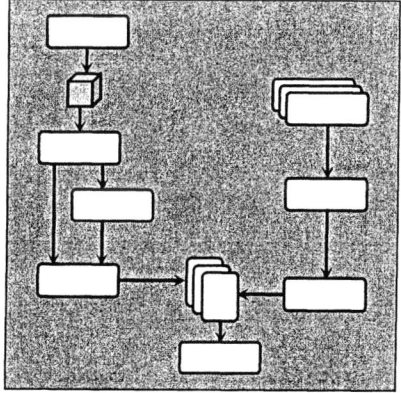

Figure 7.8 A CBLE supplies an integrative view of the organisational levels

Despite the conformance of an AMBIT type tool's functionality to the expectations of a CBLE, Potter insists that a critical component of a tool which supports a CBLE is its user friendly GUI [Potter *et al.* 1995].

7.6 Creativity

A thing constructed can only be loved after it has been constructed, but a thing created is loved before it exists.   – Gilbert Keith Chesterton

There is no standard method for determining what is a creative work, just as there is no standard technique for inducing creativity. Nonetheless, creativity tools and techniques are employed to support the initial stages of the creative process i.e. mind maps, lateral thinking, brainstorming, bi-association, janussian thinking. However, just as there is no standard technique for inducing creativity, there is as no standard definition for creativity [Dooley 2003].

Some of the theories concerning the origins of creativity, include [Farid 1993, Henry 1991, Weisberg 1986, Bowden 1990]:
- *Accident.* Creativity is a result of serendipitous good fortune.
- *Grace.* The whole notion of creativity is mysterious. It is similar Archimedes shouting Eureka, when he realised that the amount of water his body had displaced in the bath was equivalent to the loss of weight he felt of his body and that a similar principle would assist him in determining the weight of gold in the king's crown.
- *Association.* The connection of two previously unrelated concepts (e.g. X-rays and food). Creativity is that which "results in the generation of new and useful ideas or the combination of existing ideas into new and useful concepts to satisfy a need" [Farid 1993].
- *Cognitive.* There is nothing special about creativity. It is a perfectly normal cognitive process like recognition, reasoning and understanding.
- *Personality.* Only those with special abilities are capable of being creative (e.g. Mozart). Creative individuals have extraordinary personality characteristics and are thus capable of mysterious thought processes [Weisberg 1986].

Two conflicting views of the 'seeds' of creativity are the genius view and the behaviourist view. The former refers to creativity as being that which enables the owner to find significance in the irrelevant and to find meaning from contradictions. The latter perspective argues that creative responses are the result of one of two processes:
1. A 'new' situation contains elements similar to those of an old situation, thus these elements serve as the basis for generalising the old response to the new situation.
2. If the new situation is completely dissimilar to previous situations, then the solution is to behave randomly by combining various responses in numerous ways.

However, according to Malsow [1954], creativity is not just the remit of genius, but is instead the universal heritage of every human being. Hussey supports this perspective by warning organisations that in focussing on trapping genius creativity, they are "forgetting that quality of imaginative thought which is possessed to a greater or lesser extent by all human beings" [Hussey 1997]. A creative output is the result of the natural thought processes of ordinary individuals. Creativity only becomes extraordinary on the basis of what the individual produces.

7.6.1 Enhancers and Inhibitors of Creativity

From an individual perspective, one of the major obstacles to creativity is a negative attitude. Pessimists will turn their attention to the negative aspects of a problem and expend their creative energy worrying about possible unsatisfactory outcomes and dangers to his plans. Meanwhile, optimists will display creativity by concentrating on the inherent opportunities. Thus, the former, by their negative attitude suppresses creativity whilst the latter supports creative skills. Other ingredients to a non-creative state include: fear of failure, executive stress, rule following and over-reliance on logic [Goman, 1989]. Conversely, creativity is enhanced through attitude adjustment, risk-taking and stress safety valves. Creative thinking is also enhanced through the use of lateral thinking, which is the opposite of vertical thinking. Vertical thinking involves digging the same hole deeper, whilst in lateral thinking numerous new holes are dug. De Bono stated that "lateral thinking seeks to get away from the patterns that are leading one in a definite direction and to move sideways by reforming the patterns" [De Bono 1968].

During their analysis of the impediments to creativity within individuals, Woodcock *et al.* [1993] identified the following creativity barriers: personal laziness, personal habits, excessive tension, muted drive, insufficient opportunities, over seriousness and poor methodology. From an organisational perspective, barriers to creativity include [Reinartz *et al.* 2001]: intolerance of differences, overly rational thinking, inappropriate incentives and excessive bureaucracy.

> Improving the creativity of employees is important if organisations are to compete successfully in today's globally competitive environment.
>
> – Thacker 1997

7.6.2 Creative Process

Kao [1989] views the creative process of an individual as consisting of six phases with the initial phase of the creativity process being 'interest', where an individual's intuition or emotion leads him to scan his environment for opportunities or solutions. The second stage of the process is 'preparation', where planning is undertaken to 'prepare for the expedition'. Following this stage, the process enters the 'incubation' phase, where the individual utilises his intuition to 'mull things over'. This stage can be of indefinite duration and is often enhanced by outside influences and communication. The successful end of the 'incubation' phases is denoted by the 'illumination' phase, where the idea or solution finally 'comes together' from the intuition of the individual in 'the 'eureka' experience'. The next stage of the creative process is that of 'verification', where the individual rationally validates his creative output relative to the desired output. In the event that there are irregularities, the individual may return to earlier stages of the process for rework based on the new knowledge gained. In the event that the output is deemed positive, the 'exploitation' phase of Kao's model begins where the individual or organisation rationally try to "capture value from the creative act".

7.6.3 Creativity and Age

In the late 1940s a group of psychologists were discussing the lack of creativity in adults and they speculated that by the age of 45, only a minute percentage of the group would actually think creatively [Goman 1989]. In support of this, Perkins suggested that "people are better critics than creators: their knowledge is more readily available for judgement than for generation" [Perkins 1981]. However, this may be as a result of cognitive processes becoming set in 'comfort zones' and not being continuously challenged. Given an organisational culture where an individual is continuously encouraged to embrace change, it is possible that this decline in creativity may be delayed or postponed indefinitely.

A study of 738 scholars, scientists and artists born after 1600 illustrated that the productively creative years for the disparate disciplines were [Kim 1990]:

Disciplines	Ages
Artists	40s
Historians and Philosophers	60s
Technical Contributors	Span active spectrum
Biologists	40s
Geologists	50s
Inventors	60s
Mathematicians	They generally tend to reach their zenith between 30 - 45 years of age

A study of the creativity zenith in 420 literary contributors, highlighted the following [Kim 1990]:

Writers	Ages
Poets	The age for the majority of renowned work was 38.5
Writers of imaginative prose	41.6
Writers of non-fiction	50.3
Writers of informative prose	46.7

In the case of the average individual, creativity increases with age until it peaks towards the middle years and then commences its decline (Figure 7.9).

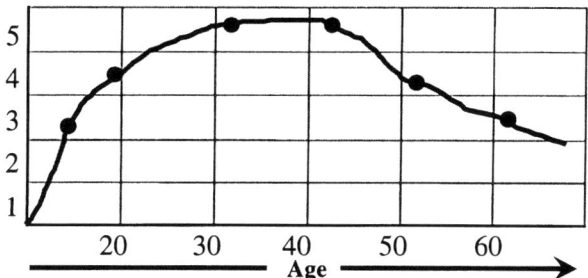

Figure 7.9 Life expectancy for creativity [Kim 1990]

Yet, how can one determine when one is being creative? Johnson-Laird suggests that an act of creation occurs when [Johnson-Laird 1993]:
- The result is formed from existing elements, but in a manner that is unique and novel, such that it is perceived as a new perspective or product.
- The result is not constructed by some simple deterministic procedure, but is instead established by enforcing freedom of choice and expression.
- The result does not depend on generalisation.

Whatever the result, creativity is an output-oriented process [Drazin *et al.* 1999], involving individuals cognitively, behaviourally and emotionally to realise creative artefacts [Kahn *et al.* 1990], in the form of original, innovative and useful ideas, products, processes or procedures [Amabile 1988, Oldham *et al.* 1996].

Hussey [1997] quotes Maslow's [1954] perspective that "the greatest contribution to creativity may be expected from people who have reached the self actualisation stage.... [and their lower] needs are already met". Thus the development of the employees and the fulfilment of their motivational needs can have a positive impact on the creative capabilities of the organisation. Notwithstanding, people have an inherent capability to be creators or inhibitors in the creative process [Berger *et al.* 1995]. This is a perspective supported by Senge [1990], who views resistance as "neither capricious nor mysterious. It almost always arises from threats to traditional norms and ways of doing things". Rather than pushing harder to overcome this resistance to change, smart leaders identify the source of the resistance and adopt a participative approach. Employees, irrespective of the management perspective, view themselves as a stakeholder within the organisational structure. Thus, there must be clear communication between the organisational layers as to the need for creativity in order to encourage all to embrace it and work towards its positive realisation.

7.6.4 Creativity and Tool Design

"If technology is to play any part at all in extending the boundaries of human thought and actions, then a critical issue is how to design the technical systems in such a way as to foster creativity" [Edmunds *et al.* 2000]. However, despite the advances in IT tools and technologies, Selker *et al.* suggest that the design of current system interfaces creates obstacles to creativity. They propose that interfaces should be designed to support and encourage human creativity or at the very least reduce the inhibitors to creativity [Selker *et al.* 2002].

Some of the ways in which the design of IT tools can be used to enhance creativity are [Candy *et al.* 2002, Shneiderman 2002, Kim 1990, Goman 1989, Elam *et al.* 1987, deBono 1968, deBono 1981]:

- *Breaking the perceptual and cognitive sets.* The cognitive and perceptual sets represents the states whereby:
  1. The learned conventions of an individual may prevent him from perceiving the world in new ways (perceptual set).
  2. A restricted range of solution approaches repeatedly used by an individual, may detract him from the consideration of new approaches and strategies (cognitive set).
- *Supporting divergent thinking.* Divergent thinking is characterised by the production of a variety of alternative solutions for a particular task.
- *Immersion.* The task is supported by the complete immersion of the user in his activity.
- *Holistic views.* The user is able to view the task from alternative perspectives.
- *Parallel channels.* Different views and approaches are made active concurrently.
- *Enhancing task motivation.* Task motivation represents an individual's attitude to a task, in combination with his perceptions of the reasons for undertaking the task.
- Providing in depth *search facilities.*
- *Visualisation of information.*
- *Exploring, reviewing and disseminating results.*
- *Supporting Incubation and Retained Control.* Incubation refers to the cessation of conscious effort on a problem for a period of time. Retained control refers to an individual's execution of the task in any manner he chooses.
- *Facilitating Competence.* Competence pertains to the reduction of criticism and the affirmation of confidence in order to encourage greater competence.
- *Reducing Stress.* Stress can often result in the early termination of alternative idea generation. As a result, it is important to reduce stress as much as possible.

> If individual creativity is to be confined within the frameworks permitted by IT, then however wonderful the technological advances, there will be a drying up of a part of the human potential.   – Lee 1998

Kim suggests that creative problem solving is supported, by encouraging the following factors of creativity [Kim 1990]: purpose, structure and representation (Figure 7.10).

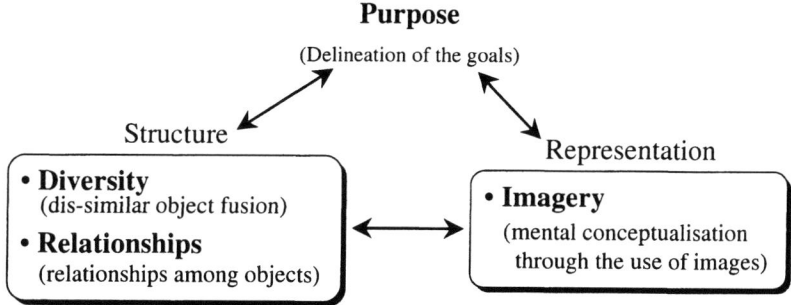

Figure 7.10 Kim's model of the factors of creativity [after Kim 1990]

Frontczak *et al.* suggest that that an understanding of learning styles could be an important factor in enhancing creativity [Frontczak *et al.* 1991].

7.6.5   Creativity as an Aspect of an AMBIT Type Tool

An AMBIT type tool should cater to enhanced creativity through the following:
- The accommodation of the learning style of the targeted user.
- Different views can be kept active concurrently.
- Exploration is supported.
- Evaluation of possible solutions is encouraged.
- The inclusion of an extensive help facility and tutorial programme to help reduce stress and increase competence.
- Experimentation is supported. Thus, the user can engage in what-if scenario building.
- The tool acts as a support suggesting options. It does not definitively propose a fixed solution.

An AMBIT based tool should promote the following skills, suggested to be vital for today's management [Alkhafaji 1995]:
- The ability to manage change and transition, by supporting experimentation.
- Proficiency in developing global strategic thinking skills and implementing ideas, through the integration of manufacturing strategy to business strategy (and ultimately to corporate strategy).
- The ability to change the way managers think and operate, as a result of an AMBIT type tool within the context of a CBLE.
- The ability to be creative and learn, by catering to enhanced creativity and accommodating the user's learning style, in the GUI design.

In Figure 7.11, Lawrie outlines the process involved in establishing a strategic direction [Lawrie 1996]. Each of these process steps; evaluation of performance and development data, generating strategic choices etc., should be catered to in an AMBIT type tool, as a result of its CBLE nature.

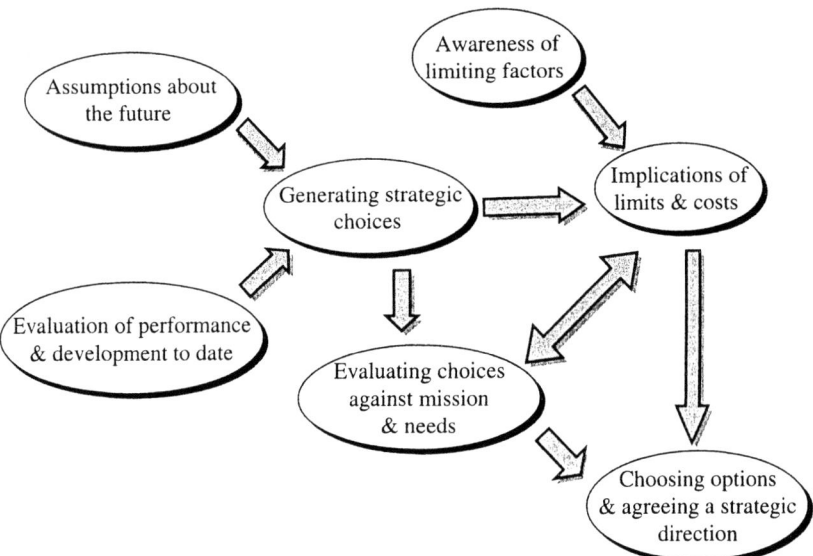

Figure 7.11 The process involved in setting a strategic direction [Lawrie 1996]

7.7   Conclusion

Design is critical. Software houses in recognising this, can spend up to as much as two thirds of their time working on the design. Yet still, they seem to "miss the mark by miles" [Constantine 1998]. Good design is difficult. It requires that the system must satisfy the following; meet the user needs, be easy to learn, easy to use and consistent. However, one aspect of design which seems to get little press is the fact that most people have strengths and weaknesses in learning. Each of us learns in different ways [Bal *et al.* 2001]. Thus, if someone is building and designing tools to assist us in the way in which we process information, it is important that learning differences are taken into account, for increasingly, the user's attitude to a system is conditioned predominantly by his experience with the system GUI. Satisfactory GUI design is crucial to system success, as the user demands software tools that are easy to use and which satisfy his needs [Rosenstein 2001, Constantine 1998, Cooper 1994, Winograd 1994, Leidner 1993]. For this reason, the four learning styles arising from the Kolbian learning circle are profiled and discussed.

A CBLE (*computer based learning environment*) essentially provides decision makers with greater scope of learning through experimentation and reflection. It assists in providing a structured way of thinking about complex problems. It also supports decision-makers in availing of opportunities, which are not easily achievable in daily management activities. The AMBIT approach as a potential CBLE was argued.

Finally, the increasing emphasis on creativity has meant that research is focussed on how to design technical systems to enhance and foster creativity. In the latter parts of this chapter, some of the inhibitors and enhancers of creativity are presented. Furthermore, some of the ways in which the design of IT tools can be used to support and foster creativity are proposed.

Bibliography and Further Reading

Aamodt, Agnar and Enric Plaza: *"Case-Based Reasoning: foundational Issues, Methodological Variations and Systems Approaches"*, AICom – Artificial Intelligence Communications, IOS Press, Volume 7, No. 1, pp. 39-59, 1994.

Alkhafaji, Abbass: *"Competitive Global Management: Principles and Strategies"*, St. Lucie Press, 1995.

Allen Dennis: *"Beyond the GUI"*, BYTE, February 1994.

Alting, L.: *"LIfe cycle design of products: A new opportunity for manufacturing enterprises"*, Concurrent engineering: Issues, Technology and Practice, Auerbach Publishers, 1991.

Amabile, T.M.: *"A Model of Creativity and Innovation in Organisations"*, Research in Organisational Behaviour, Volume 10, pp. 123-167, 1988.

Andreson, James C, H. Håkansson and J. Johanson: *"Dyadic Relationships within a Business Network Approach"*, Journal of Marketing, Vol. 58, pp 1-15, 1994.

Andrew, Charles G.: *"Product Innovation: Speeding New Ideas to Market for Competitive Advantage"*, 33rd International Conference Proceedings, American Production and Inventory Control Society, Inc., Oct. 1990.

Anonymous (1): *"Concurrent Engineering drives Product Turnaround"*, Machine Design, pp 19, UK, January 1994.

Anonymous (2): *"The Race To The Market Heats Up"*, Machine Design, pp 14, UK, 8th January 1993.

Anonymous (3): *"How to Make Concurrent Engineering Work: Part 6"*, Machine Design, pp77-79, UK, 26th November 1993.

Australian Electronics Development Centre (AEDC): *"Performance Measurement: Measuring the Performance of your Business Operations"*, 1996. http://www.aedc.com.au/PMSR.htm.

Axelsson, Björn: *"The Development of Network Research - A Question of Mobilization and Perseverance"*, In: Industrial Networks. A New View on Reality, London, UK, Routledge, 1992.

Azzone, G., C. Masella and U. Bertele: *"Design of Performance Measures for Time-based Companies"*, International Journal of Operations and Production Management, Vol. 11, No 3, pp 77-85, 1991.

Bagg, Thomas C.: *"Case-Based Reasoning Overview"*, 1997. http://hope.gsfc.nasa.gov/RECALL/homepg/cbr1.htm.

Bal, Jay and Yvette James-Gordon: *"Learning Style Preferences of Engineers in Automotive Design"*, Journal of Workplace Learning, Vol. 13, No. 6, 2001.

Baldry, D. and D. Amaratunga: *"Moving from performance measurement to performance management"*, Facilities, Volume 20, Number 5/6, pp. 217-223, 2002.

Barnatt, Christopher: *"Virtual Organisation in the Small Business Sector: The Case of Cavendish Management Resources"*, International Small Business Journal, Vol. 15, Issue 4, pp 36-47, July-September 1997.

Barnes, David: *"The complexities of the manufacturing strategy formation process in practice"*, International Journal of Operations & Production Management, Volume 22 Number 10, pp. 1090-1111, 2002.

Barnett B. J., C. J. Arbak, J. L. Olson and L. C. Walrath: *"A Framework for Design Traceability"*, Proceedings of the Human Factors Society, 36th Annual Meeting 1992.

Barth, T.S.: *"Organisational-Level Performance Measurement for External Benchmarking"*, 1996. http://technology.ksc.nasa.gov/WWWaccess/95report/ief/ie14.html.

Batra, D. and R. Santhanam: *"The Human Computer Interface in Information System Design"*, The DATA BASE for Advances in Information Systems", Winter, Volume 29, Number 1, 1998.

Bauer, Bowden, Browne, Duggan and Lyons: *"Shop Floor Control Systems: From Design to Implementation"*, Chapman and Hall, 1991.

Beaver, Diane: *"NLP for lazy learning"*, Element Books, 1998.

Becker Associates: *"Introduction to (QFD) Quality Function Deployment"*, 1997. http://www.becker-associates.com/whatsqfd.htm.

Bek Lilla and Annie Wilson: *"What Colour are you?"*, Aquarian Press, 1987.

Belson, D. and D. Nickelson: *"Measuring Concurrent Engineering Costs and Benefits"*, PED-Vol. 59, Concurrent Engineering, ASME, pp. 443-451, 1992.

Benjamins, V. R., D. Fensel and A. G. Perez: *"Knowledge Management through Ontologies"*, Second International Conference on Practical Aspects of Knowledge Management (PAKM98), Basel, Switzerland, 29-30 Oct. 1998.

Bennet, Alex: *"Lean Manufacturing"*, 1996. http://www.acq-ref.navy.mil/wcp/lm.html.

Berger, L.A., M. Sikora, and D.R. Berger: *"The Change Management Handbook: A road map to corporate transformation"*, Irwin: New York, 1995.

Berquist, Thomas and C. R. DeBiase: *"Commerce One"*, Goldman Sachs, Jan. 26, 2001.

Bigelow, John D.: *"Managerial Skills: Explorations in Practical Knowledge"*, Sage, 1991.

Biocca, Frank: *"Communication within Virtual Reality: Creating a Space for Research"*, Journal of Communication, Vol. 42, Part 4, 1992.

Birkey, R.C., Rodman, J.J.: *"Adult learning styles and preference for technology programs"*, Lifelong Learning Conference Proceedings, 1995.

Blackerby, Philip: *"Strategic Planning"*, 1996. http://www.perfstrat.com/articles/overview.html.

Blackhouse, C.J. and N.D. Burns: *"Agile Value Chains for Manufacturing – Implications for Performance Measures"*, International Journal of Agile Management Systems, Volume 1, Number 2, 1999.

Blackmoore, J.: *"Learning style preferences online"*, Telecommunications for Remote Work and Learning, 1996.

Blau, J.R.: *"Europe Turns Competitive With Concurrent Engineering"*, Machine Design, pp 41-44, June 1994.

BOMA, *"Business Object Management Architecture"*, 1995. http://www.sesh.com/essay.htm.

Boone, Mary E.: *"Leadership and the Computer"*, Prima Publishing, 1991.

Bort, Julie: *"Extended Enterprise takes Root"*, NetworkWorldFusion, 2003. http://www.nwfusion.com/ee/2003/eeopen.html.

Bourne, M, J. Mills, M. Wilcox, A. Neely and K. Platts: *"Designing, implementing and updating performance measurement systems"*, International Journal of Operations and Production Management, Volume 20, No. 7, pp. 754-771, 2000.

Bower, J.L. and T.M. Hout: *"Fast-Cycle Capability for Competitive Power"*, Harvard Business Review, November-December, pp 110-118, 1988.

Boykin, R.: *"Consortium for Advanced Manufacturing – International (CAM-I)"*, 3301 Airport Freeway, Suite 324, Bedford, TX 76021, 1997. http://www.cam-i.com.

Bradley, P.: *"A Performance Measurement Approach to the Re-engineering of Manufacturing Enterprises"*, Ph.D. thesis, CIMRU, UCG, 1996.

Bradley, Padraig, J. Browne, S. Jackson and H. Jagdev: *"Business Process Re-engineering: A study of the software tools currently available"*, Computers in Industry, Vol. 25, No. 3, pp. 309-330, March 1995.

Brand, S.C.: *"Strategic Planning in Emerging Companies"*, Addison-Wesley, 1981.

Brennan, A.: *"A Graphical User Interface Design Tool to Facilitate Managerial Learning"*, Ph.D. Thesis, CIMRU, NUI, Galway, 1996.

Brennan, B.A.: *"Hands of Light"*, Bantam Books, 1988.

Breslauer, James M.: *"Culture Change in the Organisation of the Future"*, 33rd International Conference Proceedings, American Production and Inventory Control Society, Inc., Oct. 1990.

Brindley, Keith: *"Newnes Electronics Assembly Handbook"*, Newnes, Oxford, 1993.

Brown, Hanbury: *"Man and the Stars"*, Oxford University Press, 1978.

Brown, Steve: *"Strategic Manufacturing for Competitive Advantage: Transforming Operations from Shop Floor to Strategy"*, Prentice Hall, London, 1996.

Browne, A.D., P.R. Hale and J. Parnaby: *"An Integrated Approach to Quality Engineering in Support of Design for Manufacture"*, Design for Manufacture - Strategies, Principles and Techniques, ed. John Corbett *et al.*, Addison Weseley, 1991.

Browne, J. and C. Mc Mahon: *"CAD/CAM - From Principles to Practice"*, Addison -Wesley, London, 1993.

Browne, J., John Harhen and James Shivnan: *"Production Management Systems: A CIM Perspective"*, Addison-Wesley, 1988.

Browne, J., P.J. Sackett and H. Wortmann: *"Manufacturing Business Challenges in the late 1990s"*, 1st SCMA Conference on Outstanding Success, 1994.

Browne, J., P.J. Sackett and J.C. Wortmann: *"Future Manufacturing Systems - Towards the Extended Enterprise"*, Computers in Industry, Vol. 25, pp. 235 - 254, 1995.

Browne, J., P.J. Sackett and J.C. Wortmann: *"The System of Manufacturing - A prospective Study"*, CEC DG XII, 1994.

Browne, J. and J. Zhang: *"Extended and Virtual Enterprises – Similarities and Differences"*, International Journal of Agile Management Systems, Vol. 1, No. 1, 1999.

Buch, Kim and Susan Bartley: *"Learning style and training delivery mode preference"*, Journal of Workplace Learning, Volume 14, Number 1, pp. 5-10, 2002.

Buday, Robert S.: *"Reengineering One Firm's Product Development and Another's Service Delivery"*, Planning Review, March/April 1993.

Burnes, Bernard: Managing Change: *"A Strategic Approach to Organisational Dynamics"*, Pitman Publishing, 1996.

Burns, Jack: *"CyberSex"*, Sunday World, 1993.

Burns, W.: *"Profit as a performance measure: powerful concept, insufficient measure"*, Performance Measurement - Theory and Practice: The First International Conference on Performance Measurement, Cambridge, 1998.

Buxton Bill: *"Human Input and Interaction: Improving the State of the Science of the Art"*, 1995, http://www.dgp.toronto.edu/HCI/torchi/meetSummaries/Nov.

Candy, Linda and Ernest Edmonds: *"Creativity, art practice and knowledge"*, Communications of the ACM, Volume 45, Number 10, 2002.

Canty, E.J.: *"Simultaneous Engineering: Expanding Scope of Quality responsibility"*, Digital Equipment Corporation White Paper, 1987.

Capra, Fritjof: *"The Turning Point"*, London, Fontana, 1983.

Carlisle, J.A. and R.C Parker: *"Beyond Negotiation"*, Chichester, John Wiley & Sons, 1989.

Carlson, Susan, Mark Baginski and Natasha Ter-Minassian: *"A Passport to Integrated Design: Implementing Concurrent Engineering in a Small to Medium-Sized Company"*, 1996. http://vlead.mech.virginia.edu/publications/dtm96/dtm96.html.

Carlsson, M.: *"Product development - concepts and strategies for management of product development; the cases of successful Swedish manufacturing industries, in Managing human resources in R&D"*, R&D Management Conference, 4-6 July 1994, Manchester.

Cavinato, J.: *"Evolving Procurement Organisations: Logistic Implications"*, Journal of Business Logistics, Vol.13, No.1, 1991.

Chan, Felix T. S. and H.J. Oi: *"Feasibility of performance measurement system for supply chain: a process-based approach and measures"*, Integrated Manufacturing Systems, Volume 14, Number 3, pp. 179-190, 2003.

Cheng, Eddie E. L., Heng Li and Danny C. K. Ho: *"Analytic hierarchy process (AHP): A defective tool when used improperly"*, Measuring Business Excellence, Volume 6, Number 3, pp. 33-37, 2002.

Chijiiwa Hideaki: Colour Harmony: *"A Guide to Creative Colour Combination"*, Rockport Publishers, 1987.

Childe, S.J.: *"The extended enterprise - a concept of cooperation"*, Production, Planning and Control, Vol. 9, No. 4, 1998.

Chira, Ovidu, C. Chira, T. Roche, D. Tormey and A. Brennan: *"An agent-based approach to knowledge management in distributed design"*, awaiting publication, 2003.

Choy, K. L. and W. B. Lee: *"A generic tool for the selection and management of supplier relationships in an outsourced manufacturing environment: the application of case based reasoning"*, Logistics Information Management, Volume 15, Number 4, pp. 235-253, 2002.

Christense, Soren and Lone Dybkjaer: *"The Global Short-Circuit and the Explosion of Information"*, 1996. http://www.fsk.dk/fsk/pub1/info2000-uk/chap01.htm.

Chu, William W., Wan Ching Lin, Van Le, Maureen Weicher and Dominic Yu.: *"Business Process Re-engineering Analysis and Recommendations, Management"*, p.14, November 1994.

Ciampa, D.: *"Total Quality"*, Addison-Wesley, Reading, MA, 1992.

Ciampa, Dan.: *"Reengineering with Caution"*, Management Review, October 1993.

CIMLINC Inc.: *"Lean Manufacturing - A Two Legged Stool Won't Work"*, 1997. http://www.cimlinc.com/msg/e5story.htm.

Clark Colvin Ruth: *"Authorware, Multimedia and Instructional Methods"*, Macromedia Authorware, Taking the Plnge, Macromedia Inc., 1995.

Clark, K.B. and T. Fujimoto: *"Product Development Performance: Strategy, Organisation and Management in the World Auto Industry"*, Harvard Business School Press, Boston, 1991.

Clark, Michael P.: *"Stop Being the Boss and Start Being a Manager"*, 33rd International Conference Proceedings, American Production and Inventory Control Society, Inc., Oct.1990.

Classe, A.: *"Software Tools for Re-engineering"*, Business Intelligence, London, 1994.

Cleetus, K.J.: *"Virtual Team Framework and Support Technology"*, NATO AI Series F: Computer and Systems Science CE: Tools and Technologies for Mechanical Design, Vol. 108, pp 41-73, 1993.

Clemons, Eric. K. and Paul R. Kleindorfer: *"An economic analysis of interorganisational information technology"*, Decision Support Systems, Issue 8, pp 431-446, 1992.

Connor S.J.: *"Motivating the Individual Manager to Meet Enterprise Goals"*, APICS Conference Proceedings, 1990.

Consortium for Advanced Manufacturing, *"Next Generation Manufacturing Systems"*, http://www.cam-i.org/ngms.html.

Constantine, Larry L.: *"What do users want Engineering Usability into Software"*, Windows Tech Journal, 1998.

Cooper, Alam: *"In my Opinion"*, Common Sense, 1994. http://www-pcd.stanford.edu/asd/info/articles/common-sense.html.

Cotta-Schønberg, Michael v.: *"Performance Measurement in the context of Quality management"*, 1st Northumbria Conference on Performance Measurement, Copenhagen Business School, 1995. http://sun1.lib.cbs.dk/people/mcs/perform.html.

Coughlan, P.D. and A. R. Wood: *"Developing Manufacturable New Products"*, Business Quarterly, Summer, pp. 49-53, 1991.

Croft, Chris: *"Engineer Management 333 Lectures"*, 1995.
http://www.ee/uwa.eu/~ccroft/em333/lec.thml.

Currie, Melanie: *"The Learning Organisation - What Role For IT?"*, 1997.
http://www.nbs.ntu.ac.uk/staff/currimj/it_issue/old/learning.htm.

Davenport, T.H. and J.E. Short: *"The New Industrial Engineering: Information Technology and Business Process Redesign"*, Sloan Management Review, pp. 11 – 27, Summer 1990.

Davenport, Thomas H.: *"Process Innovation: Reengineering Work through Information Technology"*, Ernst and Young, 1993.

Davidow, William H. and Michael S. Malone: *"The Virtual Corporation: Structuring and Revitalizing the Corporation for the 21st Century"*, New York: HarperCollins Publishers, 1992.

Davidson, W. H.: *"Beyond Re-Engineering: The Three Phases of Business"*, Transformation', IBM Systems Journal, Volume 32, No. 1, pp. 65 - 79, 1993.

Davis, Roger A.: *"CIM Strategy Development and Justification"*, 33rd International Conference Proceedings, American Production and Inventory Control Society, Inc., Oct. 1990.

de Bono, Edward: *"Atlas of Management Thinking"*, Penguin Books, 1981.

de Bono, Edward: "New Think: The Use of Lateral Thinking in the Generation of New Ideas", NY, Basic, 1968.

Dean, Edwin B.: *"Design for Value: From the perspective of competitive advantage"*, 1996. http://akao.larc.nasa.gov/dfc/dfval.html.

Deise, M., Novikow, C., King, P. and Wright, A.: *"Executives Guide to E-Business - From Tactics to Strategy"*. New York: John Wiley & Sons, Inc. 2000.

Dewhurst, Frank W., Angel Rafael Martínez-Lorente, Cristóbal Sánchez-Rodríguez: *"An initial assessment of the influence of IT on TQM: a multiple case study"*, International Journal of Operations and Production Management Volume 23, Number 4, pp. 348-374, 2003.

DFMA, http://www.dfma.com.

Dixon, J.R., A.J. Nanni and T.E. Vollmann: *"The New Performance Challenge"*, Business One Irwin, Homewood Illinois, 1990.

Dixon, N. M. and M. D. Svinicki: *"The Kolb model modified for classroom activities"*, College Teaching, Volume 35, Number 4, pp. 141-146, 1995.

Dobrzeniecki, Aimee: *"Executive Information Systems"*, 1994.
http://mtiac.hq.iitri.com/MTIAC/pubs/eis/eis.

Dobyns, Lloyd and Clare Crawford: *"The Global Marketplace"*, from Candace Deans and Jaak Jurison (eds.), *"Information Technology in a Global Business Environment"*, Readings and Case's, Boyd and Fraser publishing, 1996.

Dolan, Thomas: *"An Analysis of the Strategic Implications of Interconnect Technologies"*, M. Eng. Sc. Thesis, CIMRU, NUI, Galway, 1992.

Dooley, Lawrence: *"A Study of Radical Change Programmes in Irish Industry"*, Masters Thesis, Commerce department, NUI, Galway, 1997.

Dooley, Lawrence and A. Brennan: *"Networked Creativity: An Essential Competitive Capability"*, awaiting publication, 2003.

Drazin, R., M.A. Glynn and R. K. Kazanijan: *"Multilevel theorizing about creativity in organisations: A sensemaking perspective"*, Academy of Management Review, April, pp. 286-307, 1999.

Drucker, P. F.: *"Post Capitalist Society"*, Harper Business, 1994.

DTI (Department of Trade and Industry, UK): *"The Government's Manufacturing Strategy"*, 2002. http://www.dti.gov.uk/manufacturing/strategy.pdf.

Dumas, Joseph and Paige Parsons: *"Discovering the Way Programmers think about New Programming Environments"*, Communications of the ACM: Cognition and Software Development, Volume 38, No. 6, June 1995.

Dunk, Alan S.: *"Product quality, environmental accounting and quality performance"*, Accounting, Auditing & Accountability Journal, Volume 15, Number 5, pp. 719-732, 2002.

Dunne, K., J. Browne and H.S. Jagdev: *"RapidPDM: Faster implementation of PDM systems"*, PDT Europe 2001 - Annual conference on Product Data Technology, Brussels, Belgium, 24-26 April 2001.

Eakin, W. K., Gardiner and Jawaharnayak: *"Cost and Performance Criteria for Selection of Materials and Processes in Micro-electronics Packaging"*, Journal of Electronics Manufacturing, Vol. 1, pp. 13 - 22, 1991.

Edmunds, E. and L. Candy: *"Creativity, Art Practice and Knowledge"*, Communications of the AMC, Volume 45, Number 10, 2002.

Edwards, Philip R.: *"Manufacturing Technology in the Electronics Industry"*, Chapman and Hall, 1991.

Elam, Joyce J and Melissa Mead: *"Designing for Creativity: Considerations for DSS Development"*, Information and Management, Vol. 13, pp. 215 - 222, 1987.

ELANCE: *"Infrastructure for the Extended Enterprise : How Services Procurement, and Management enables the extended enterprise"*, 2002. http://images. elance.com/images/Elance_InfrastructurrfortheExtendedEnterprise.pdf.

Eloranta, Eero, J. Nikkola, J. Ranta and M Ollus: *"Challenges for New Manufacturing Paradigms - Towards concurrency via Time Based Competition"*, In: Storch, Richard L. (ed.), Proceedings of the International IFIP WG5.7 Working Conference, September 11-15, Seattle, WA, USA, 1995.

Elshaug Lene, Sven Kruse-Jensen and Anders Kjellsen, *"Decision Making"*, 1995. http://www.pakt.unit.no/~alk/ak/decision.htm#innlednin

ENAPS - The European Network for Advanced Performance Studies (ESPRIT Project no. 20888), *Final Report* 7/1999. http://www.enaps.com

Evans, Philip B. and Thomas S. Wurster: *"Strategy and the New Economics of Information"*, Harvard Business Review, September-October, pp 70-82, 1997.

Everaert, Patrica: *"Cost targets and time pressure during new product development"*, International Journal of Operations & Production Management, Volume 22, Number 12, pp. 1339-1353, 2002.

Eversheim, W. and M. Gross: *"Trends and Experiences in Applying Simultaneous Engineering"*, 1991.

Feigenbaum, Edward, Pamela McCorduck and H. Penny Nii: *"The Rise of the Expert Company: How visionary companies are using Artificial Intelligence to achieve higher productivity and profits"*, Vintage Books, 1989.

Felder, R.M.: *"Matters of Style"*, ASEE Prism, Vol. 6, Number 4, pp. 18-23, 1996.

Fiedler, Fred: "Harnessing Managerial Creativity and Intellectual Resources for Effective Management", 23rd International Congress of Applied Psychology, Madrid, 1994.

Finke, Ronald A., Thomas B. Ward and Steven M. Smith: *"Creative Cognition: Theory, Research and Applications"*, MIT Press, 1993.

Forbairt: *"Virtual Corporation Defined"*, Summary Section for Forbairt Internet Report, Forbairt, Ireland, 1996.

Fraser, Julie: *"Customers Drive Desires Home: How to Synchronise when Mass Customisation comes to Roost"*, Intelligent Manufacturing, June 1995.

Freeman, C.: *"The Factory of the Future: The Productivity Paradox, Japanese Just-In-Time and Information Technology"*, ESRC PICT Policy Research Paper 3, 1988.

Frontczak Nancy T., and Lexis F. Higgins: *"Learning Styles and Creativity Training: Implications for Information Management"*, Proceedings of the 24th Annual Hawaii International Conference on System Sciences, IEEE Computer Society Press, 1991.

Gaarder, Jostein: *"Sophie's World"*, Phoenix House, Great Britain, 1995.

Gates, B.: *"Business @ the Speed of thought"*. New York: Warner Books, 2000.

Ginsberg, Gerald L.: *"Electronic Equipment Packaging Technology"*, Van Nostrand Reinhold, New York, 1992.

Gloor, P: *"Making the e-business Transformation"*, Springer-Verlag London Berlin Heidelberg, 2000.

Goggin, K., P. Jordan, and J. Browne: *"Resource Sustainment- Issues and Opportunities"*, ME-SELA 97, Loughborough University, 1997.

Goldman Steve and Kenneth Preiss (editors): *"21st Century Manufacturing Enterprise Strategy: Infrastructure"*, Iacocca Institute, Vol. 1, Nov. 1991.

Goman, C.K.: *"Creative Thinking in Business - A Practical Guide"*, Kogan Page 1989.

Goodman, Richard A. and Michael W. Lawless: *"Technology and Strategy - Conceptual Models and Diagnostics"*, Oxford University Press, 1994.

Gott, B.: *"Empowered Engineering for the Extended Enterprise – A Management Guide"*, Cambashi Ltd., Cambridge, England, 1996.

Graham, G. and Pervaiz Ahmed: *"Buyer-supplier management in the aerospace value chain"*, Integrated Manufacturing Systems, Volume 11, Number 7, pp. 462-468, 2000.

Grunberg, Thomas: *"A review of improvement methods in manufacturing operations"*, Work Study, Volume 52, Number 2, pp. 89-93, 2003.

Håkansson, H and I. Snehotna: *"No Business is an Island: The Network Concept of Business Strategy"*, Scandinavian Journal of Management, Vol.4, Issue. 3, pp 187-200, 1989.

Håkansson, Håkan and Ivan Snehotna (eds): *"Developing Relationships in Business Networks"*, Routledge, London, 1995.

Hamaker, Jamie L. And Marc J. Schniederjans: *"A new strategic information technology investment model"*, Management Decision, Volume 41, Number 1, pp. 8-17, 2003.

Hamel, G. and C. Prahalad: *"The Core Competence of the Corporation"*, Harvard Business Review, May-June, pp 26-29, 1990.

Hamilton, Joan O, Emily T. Smith, Gary Mc Williams, Evan I. Schwartz and Joan Carey: *"Virtual Reality - How a Computer Generated World could Change the Real World"*, Business Week, Oct. 1992.

Hammer, M. and J. Champy: *"Reengineering the Corporation: A Manifesto for Business Revolution"*, Harper Business press, New York, 1993.

Hammer, Michael: *"Reengineering Work: Don't Automate, Obliterate"*, Harvard Business Review, July-August 1990.

Handy, Charles: "Beyond *Certainty: The Changing World of Organisations"*, Hutchinson 1985.

Handy, C.: *"The Age of Unreason"*, Arrow: London, 1989.

Handy, Charles: *"Gods of Management"*, Pan London, 1985, quoted in Bigelow's Managerial Skills Explorations in Practical Knowledge, Sage, 1991.

Hanka, Rudolf and Karel Fuka: *"Information overload and just-in-time knowledge"*, The electronic library, Vol. 18, No. 4, 2000.

Harbi, S., R. Calvi, and M. Le Dain: *"Managing Early Supplier Involvement (ESI) into the New Product Development Process (NPDP), a case study"*, in: Weber, F. and K.S. Pawar (Eds.), Proceedings of the 6th. International Conference on Concurrent Enterprising, Toulouse, France, June 2000.

Harris, Philip R.: *"European challenge: developing global organisations"*, European Business Review, Volume 14, Number 6, pp. 416-425, 2002.

Harrison, E. Frank and Monique E. Pelletier: *"Revisiting Strategic Decision Success"*, Management Decision, Volume 39, Number 3, pp. 169-180, 2001.

Hartwick Jon and Jenri Barki: *"Explaining the Role of User Participation in Information System Use"*, Management Science, Vol. 40, No. 4, April 1994.

Hardwick, M., D.L. Spooner, T. Rando and K.C. Morris: *"Sharing Manufacturing Information in Virtual Enterprises"*, Communications of the ACM, February, Vol. 39, No. 2, pp 46-54, 1996.

Haque, Badr: *"Problems in concurrent new product development: an in-depth comparative study of three companies"*, Integrated Manufacturing Systems, Volume 14, Number 3, pp. 191-207, 2003.

Hauser, J.R. and D. Clausing: *"The House of Quality"*, Harvard Business Review, May-June, pp. 63-73, 1988.

Hayes R.H., and S.C. Wheelwright: *"Restoring our Competitive Edge, Competing through Manufacturing"*, Wiley and Sons Inc., 1984.

Hayes, R. and Pisano G.: *"Beyond World-Class: The New Manufacturing Strategy"*, Harvard Business Review, January-February, pp. 77 - 86, 1994.

Hayes, R. and S. Wheelwright: *"Revolutionising Product Development"*, Free Press, New York, 1992.

Hayes, R.H., S.C. Wheelwright and K.B. Clark: *"Dynamic Manufacturing - Creating the Learning Organisation"*, The Free Press, 1988.

Hayes, T.: *"Is Radio sounding the end of the TV Age?"*, Irish Independent, Friday, August 6th, 1993.

Haynes I and N. Frost: *"Accelerated Product Development: An Experience with Small and Medium Sized Companies"*, World Class Design to Manufacture, pp. 32-37, Vol. 1, No. 5, 1994.

Hebblethwaite Jeremy: *"Learning Organisations"*, 1996.
http://www.ee.ed.ac.uk/~MEAB/learning_organisation/building_blocks.html.

Hedelin, Lisbeth and Carl Martin Allwood: *"IT and Strategic Decision Making"*, Industrial Management and Data Systems, Volume 102, Number 3, pp. 125-139, 2002.

Hefley, William: *"Helping Users Help Themselves"*, IEEE Software, 1995.

Henry, Jane: *"Creative Management"*, edited at the Open Business School, Sage Publications, 1991.

Herbig, Paul A., and Hugh Kramer: *"The effect of information overload on the innovation choice process"*, Journal of consumer marketing, Vol. 11, No. 2, 1994.

Hewlett-Packard *Quality Manual*, 1990.

Higgins, Paul: *"Master Production Scheduling: A Key Node in an Integrated Approach to Production Management Systems"*, Ph.D. Thesis, CIMRU, NUI, Galway, 1991.

Hirsch, B., G. Windhoff, J. Hoheisel, W. Echelmeyer, T. Lien and E. Morales: *"Experience with a Joint University Course Based on Distributed Simulation Games for Practising Co-operative Work and Concurrent Engineering"*, in: Weber, F. and K.S. Pawar (Eds.), Proceedings of the 6th. International Conference on Concurrent Enterprising, Toulouse, France, 28 - 30 June 2000.

Hoadley. Ellen D.: *"The Functions of Colour in Human Information Processing"*, 1989. http://attanze.loyola.edu/attanze/research/wp0989.003.html.

Hobbs, Reginald: *"Concurrent Engineering Resources on the Web"*: What is Concurrent Engineering? 1996.
http://www.cc.gatech.edu/computing/SW_Eng/people/Phd/ce

Hodgkinson, Myra: *"A shared strategic vision: dream or reality?"*, The Learning Organisation, Volume 9, Number 2, pp. 89-95, 2002.

Holtham Clive (editor): *"Executive Information Systems and Decision Support"*, Chapman and Hall, 1992.

Honey, Peter and Alan Mumford: *"The Manual of Learning Styles"*, Peter Honey, Berkshire, 1992.

Horowitz, Samuel J. and Steven K. Ladd: *"Alliance formed to promote Ceramic Interconnect Technolology"*, 1995.
http://www.dupont.com/techpaper/alliance.html.

Horton, William: *"Illustrating Computer Documentation: The Art of Presenting Information Graphically on Paper and Online"*, John Wiley & Sons, Inc., 1991.

Huang, C.Q. and K.L. Mak: *"Current Practices of Engineering Change Management in UK Manufacturing Industries"*, International Journal of Operations and Production Management, Vol. 19, No. 1, 2000.

Hubka, Vladimir: *"Theory of Technical Systems"*, Springer Verlag, 1988.

Hunt, Ingrid, J. Browne, and J. Zhang: *"The Extended Enterprise, Handbook of Life Cycle Engineering Concepts, Tools and Techniques"*, edited by Arturo Molina Gutirrez, Josef M. Senchez, Andrew Kusiak, Chapman and Hall, 1997.

Hunter, I., D. Hassan, A. Gayoso and F. Garas: *"The eLSEwise vision, development routes and recommendations"*, Engineering Construction and Architectural Management, March, 1999.

Hussey, D.: *"The Innovation Challenge"*, Wiley, Chichester, UK, 1997.

Iacocca Institute: *"21ˢᵗ Century Manufacturing Enterprise Strategy – An Industry-Led View"*, Volume 1, 1991.

Isaacs William and Peter Senge: *"Overcoming limits to learning in computer-based learning environments"*, European Journal of Operations Research 59, pp. 183-196, 1992.

IEE: http://www.iee.org.uk/PG/I7/glossary.htm.

IMS: Intelligent Manufacturing Systems, 1996, http://www.img.org.

Jackson, S.: *"Qualitative Modelling of Unstructured Knowledge to Support Strategy Determination"*, Ph.D. Thesis, CIMRU, NUI, Galway, 1991.

Jagdev, H. and J. Browne: *"The Extended Enterprise - A Context for Manufacturing"*, Invited Paper, Production Planning and Control, Volume 9, Number 3, pp 216-229, 1998.

Jagdev, H.S. and K.-D. Thoben: *"Typological Issues in Industrial Collaboration"*, Proceedings of the 5ᵗʰ International Conference on Concurrent Enterprising (ICE 1999), The Hague, the Netherlands, 15-17 March 1999.

Jagdev, H.S., J. Eschenbächer and K.-D., Thoben: *"Extended Products: Evolving Traditional Product Concepts"*, Proceedings of the 6ᵗʰ International Conference on Concurrent Enterprising (ICE2001), Bremen, Germany, 27-29 June 2001.

Jagdev, H.S. and K.-D. Thoben: *"Typological Issues of Enterprise Networks"*, International Journal of Production Planning and Control, Volume 12, No. 5, July 2001, 421-436.

Jagdev, H.S. and K.-D. Thoben: *"Anatomy of Enterprise Collaborations"*, International Journal of Production Planning and Control, Volume 12, No. 5, July 2001, 437-451.

Johnson, H. Thomas and Robert S. Kaplan: *"Relevance Lost: The Rise and Fall of Management Accounting"*, Harvard Business School Press, 1987.

Johnson-Laird, Philip N.: *"Human and Machine Thinking"*, Lawrence Erlbaum, 1993.

Jones, C., W.S. Hesterly and S.P. Borgatti: *"A General Theory of Network Goverance: Exchange Conditions and Social Mechanisms"*, Academy of Management Review, 22, 1997.

Jones, Daniel T.: *"Beyond the Toyota Production System: The Era of Lean Production"*, in C. A. Voss (ed.), Manufacturing Strategy and Process Content, Chapman and Hall, pp. 353 - 370, 1992.

Juster, N.P.: *"Traditional V Concurrent Engineering"*, 1996. http://leva.leeds.ac.uk/5330/notesW-1.html.

Kakabadse, Andrew, Siobhan Alderson and Liam Gorman: *"Irish Top Management Competencies and Team Dynamics"*, from Strategy and General Management: An Irish Reader, edited by James S. Walsh and Brian Levy, Oak Tree Press, Dublin, 1995.

Kalakota, R. and Robinson, M.: *"E-business 2.0 - Roadmap for Success"*, Upper Saddle River: Addison Wesley, 2001.

Kamath, R.R. and J.K. Liker: *"A Second Look at Japanese Product Development"*, Harvard Business Review, October-November, 1994, pp 154-170, 1994.

Kanter, Rosabeth Moss: *"The Change Masters: Corporate Entrepreneurs at Work"*, Routledge, 1995.

Kao, J.: *"Entrepreneurship, Creativity and Organisation"*, Prentice Hall, New Jersey, 1989.

Kaplan, R.S. and D.P. Norton: *"The balanced scorecard: Measures that drive performance"*, Harvard Business Review, pp. 71 – 79, 1992.

Karuppan, Corinne M.: *"Web based teaching materials: a user's profile"*, Internet Research: Electronic Networking Applications and Policy, Vol. 11, Issue 2, 2001.

Kellock, B.: *"From A to Zeta... Using Simultaneous Engineering"*, Machinery and Production Engineering, pp 22-24, UK, October 1992.

Kelly, Brian: *"An Analysis Tool to Support Strategic Manufacturing Decision Making"*, Master Thesis, CIMRU, UCG, 1995.

Kennerley, Mike and Andy Neely: *"A framework of the factors affecting the evolution of performance measurement systems"*, International Journal of Operations and Production Management, Volume 22, Number 11, pp. 1222-1245, 2002.

Kahn, W.A.: *"Psychological conditions of personal engagement and disengagement at work"*, Academy of Management Journal, Volume 33, pp. 69 – 724, 1990.

Kidd, P.T.: *"Agile Manufacturing: A Strategy for the 21st Century"*, IEEE Colloquium on Agile Manufacturing, October 26, Warwick, UK, 1995.

Kidd, P.T.: *"Agile Manufacturing: Forging New Frontiers"*, Addison-Wesley Publishing Company, Inc., 1994.

Kim, Scott: *"Interdisciplinary Co-Operation,"* The Art of Human-Computer Interface Design, Addison-Wesley Publishing Company, 1990.

Kirkwood, Craig W.: *"Strategic Decision Making: Multi-objective Decision Analysis with Spreadsheets"*, Duxbury Press, Belmont, CA, 1997.

Kolb, David: *"Experiential Learning: Experience as the Source of Learning and Development"*, Prentice-Hall, 1984.

Kolb, David, Stuart Lublin, Juliann Spoth and Richard Baker: *"Strategic Management Development: Experiential Learning and Managerial Competencies"*, Journal of Management Development, Vol. 3, Part 5, pp. 13-24, 1986.

Kolodner, Janet: *"Case-Based Reasoning"*, Morgan Kaufmann, 1993.

Konsynski, B.R.: *"Strategic Control in the Extended Enterprise"*, IBM Systems Journal, Vol.32, Issue 1, pp 111-142, 1993.

Krahe, V.A.: *"The Shape of the Container"*, Adult Learning, Volume 4, Number 4, pp 17-18, 1993.

Kueng, P.: *"Process performance measurement system: a tool to support process-based organisations"*, Total Quality Management, 11, 1, pp. 67-85, 2000.

Kurakawa, K., T. Kiriyama and Y. Baba: *"The Green Browser: An Internet-based information sharing tool for product life cycle design"*, 1997. http://green.race.u-tokyo.ac.jp/papers/cirp97/cirp97camera.htm.

Kyng, Morten: *Designing for Co-operation: "Co-operating in Design"*, Communications of the ACM, Vol. 34, No. 12, 1991.

Lacy, Marie Louise: *"Know Yourself Through Colour"*, Aquarian Press, 1989.

Lamming, R.: *"Beyond Partnership: Strategies for Innovation and Lean Supply"*, Prentice Hall, London, 1993.

Lang, Ian: *"Information Society Initiative: Foreword"*, 1996. http://www.isi.gov.uk/introisi/foreword.html.

Langley, A.: *"The Roles of Formal Strategic Planning, Long Range Planning"*, Volume 21, No. 3, pp. 40-50, 1988.

Lawrence, L., D. Andrews, B. Ralph and C. France: *"Identifying and assessing environmental impacts: investigating ISO 14001 approaches"*, The TQM Magazine, Volume 14, Number 1, pp. 43-50, 2002.

Lawrie, Alan: *"The Complete Guide to Business and Strategic Planning: for Voluntary Organisations"*, Directory of Social Change, 1996.

Lee, Choong Y: *"Total manufacturing information system: a conceptual model of a strategic tool for competitive advantage"*, Integrated Manufacturing Systems, Volume 14, Number 2, pp. 114-122, 2003.

Lee, Joe: *"Ireland the poorer for the passing of a master"*, The Sunday Tribune, February 22nd, 1998.

Lee, S. and M. Schniederjans: *"Operations Management"*, Houghton Mifflin, Boston MA, 1994.

Lefrancois, Guy R.: *"Theories of Human Learning"*, Wadworth, 4th edition, 2000.

Leidner Dorothy E., and Joyce E. Elam: *"Executive Information Systems: Their Impact on Executive Decision Making"*, Transactions of the IEEE, 1993.

Lettice, F.E.: Concurrent Engineering: *"A Team-Based Approach to rapid Implementation"*, Ph.D. Thesis, Cranfield University, UK, 1995.

Levine H. and H. Rheingold: *"The Cognitive Connection"*, New York, Prentice Hall Press, 1987.

Levy, Steven: *"Dilbert Gets Down to Business"*, The Readers Digest, August, 1997.

Line, Maurice: *"Management musings 11: measurement vs. performance"*, Library Management, Volume 24, Number 3, pp. 164-165, 2003.

Ligus, Richard G.: *"Methods to Help Reengineer your Company for Improved Agility"*, Industrial Engineering, January 1993.

Lissack, M. and J. Roos: *"Be coherent, not visionary"*, Long Range Planning, Volume 34, Number 1, pp. 53-70, 2001.

Lo, V.H.Y., P. Humphreys and D. Sculli: *"The definition method zero applied to ISO 9000 quality manuals"*, The TQM Magazine, Volume 13, Number 2, pp. 105-111, 2001.

Lowenthal, Paul, Douglas Maxwell and Tim McAloone: *"Concurrent Engineering Literature Review"*, AMBITE document, AMBITE/CU/WP2/1/01/I/P/1.0/D, 1995.

Lynch, Patrick J.: *"Visual Design for the User Interface"*, Journal of Bio-communications, 21 (1), pp. 22 - 30, 2002. http://pandora.med.yale.edu/caim/GUI.

Ma, Hao: *"To win without fighting: an integrative framework"*, Management Decision, Volume 41, Number 1, pp. 72-84, 2003

Malhotra, Yogesh: *"Business Process Reengineering and Innovation"*, 1997. http://www.brint.com/BPR.htm.

Mannion, Lorcan: *"Concurrent Engineering and Product Modelling"*, AMBITE document, AMBITE/UCG/WP2/1/01/I/P/1.0/F, 1994.

MAPICS: *"The 7 keys to World Class Manufacturing: Creating World Class Manufacturers"*, 2002.
http://www.mapics.com/downloads/documents/MPWPKEY.pdf.

Market Research Institute: *The Johnson Graduate School of Management Report on the Science Gap in Corporate Management,* October 1995.
http://www.gsm.cornell.edu/Catalog/SciMBASurvey/surveyresults.

Martin C.J.: *"Management Computer Support: Analysing the Top Managers Perspective on Interactive Systems"*, Knowledge-Based Management Support Systems, G. I Doukidis, F. Land and G. Miller (eds.), Chichester: Ellis Horwood.

Maskell, B.H.: *"Performance Measurement for World Class Manufacturing"*, Productivity Press, Cambridge, MA, 1991.

Mason-Jones, R. and D.R Towill: *"Total cycle time compression and the agile supply chain"*, International Journal of Production Economics, No. 62, 1999.

Mason-Jones, R., B. Naylor and D.R Towill: *"Engineering the Leagile Supply Chain"*, International Journal of Agile Management Systems, Vol. 2, No. 1, 2000.

McCarthy, B.: *"Creative Lesson Plans for Teaching to Learning Styles with Right/Left Mode Techniques"*, Excel, Oak Brook, IL, 1980.

McCarthy, Ian and Christos Tsinopoulos: *"Strategies for agility: an evolutionary and configurational approach"*, Integrated Manufacturing Systems, Volume 14, Number 2, pp. 103-113, 2003.

McDonald, Malcolm H.B.: *"Ten Barriers to Marketing Planning"*, Marketing Strategy, edited by Dale Littler and Dominic Wilson, Butterworth-Heinemann Ltd., 1995.

McGrew, Anthony: *"The Political Dynamics of the New Environmentalism"*, Business and the Environment: Implications of the New Environmentalism, Denis Smith (ed.), Paul Chapman publishing ltd., 1993.

McNeilly, Mark: *"Gathering Information for Strategic Decisions Routinely"*, Strategy and Leadership, Volume 30, Number 5, pp. 29-34, 2002.

Meador, C. Lawrence, Martin J. Guyote and William T. Rosenfeld: *"Decision Support Planning and Analysis: The Problems of getting large-scale DSS started"*, 1995. http://www.mstnet.com/MST/wp_ds.

Melgoza, Pauline, Pamela A. Mennel and Suzanne D. Gyeszly: *"Information Overload"*, Collection Building, Vol. 21, No. 1, 2002.

Mellor, Robert and Pavan Gupta: *"Comparing the manufacturing strategies of Australian firms with their European counterparts"*, International Journal of Operations & Production Management, Volume 22, Number 12, pp. 1411-1428, 2002.

Meyer, C.: *"How the right measures help teams excel"*, Harvard Business review, pp. 95-103, May-June, 1994.

Michels, Sjoerd: *"Human Computer Interaction"*, 1995.
http://www.eos.kub.nl:2080/w3thesis/Hci/hci.

Mill, H. and B. Ion: *"Implementing a New Design Process"*, World Class Design to Manufacture, pp. 9 - 12, Vol. 1, No. 5, 1994.

Mintzberg, Henry: *"Planning on the Left Side and Managing on the Right"*, Harvard Business Review, Vol. 54, pp. 49-58, July - August 1976.

Mintzberg, Henry: *"The Rise and Fall of Strategic Planning"*, Prentice Hall, 1994.

Mohd, Sahar Sauian: *"Labour productivity: an important business strategy in manufacturing"*, Integrated Manufacturing Systems, Volume 13, Number 6, pp. 435-438, 2002.

Møller, C.: *"Communication in Inter-organisational Operations – towards a research framework in Logistics and Production Management"*, APMS International Conference on Advances in Production Management Systems, November, Japan, 1996.

Molloy, Eoin: *"A Design Environment for Concurrent Engineering"*, Ph.D. Thesis, CIMRU, NUI, Galway, 1995.

Monaghan, Garrett: *"A Model to Support Decision Making in the Interconnect Technology Domain"*, M. Eng. Sc. Thesis, CIMRU, NUI, Galway, 1995.

Mooney, Paul: *"Developing the High Performance Organisation"*, Oak Tree Press, 1996.

Moore, John: *"The Information Age"*, 1997.
http://www.dist.gov.au/growth/html/it.html.

Morgan, Konrad, Robert L. Morris, Hamish MacLeod, and Shirley Gibbs: *"Gender Differences and Cognitive Style in Human-Computer Interactions"*, International HCI Conference, 1992.

Morgan, Gareth, Emerging Waves and Challenges: *"The Need for New Competencies and Mindsets"*, Creative Management, edited by Jane Henry at the Open Business School, Sage Publications, 1991.

Morris, Bonnie: *"Case-Based Reasoning"*, 1995.
http://www.bus.orst.edu/faculty/brownc/aies/news-let/fall95/casebase.html.

Motter-Hodgson, M.: *"Meeting the Needs of Diverse Types of Learners"*, http://www. cybercorp.net/gymv/bulla/b_v01n01.html#learn, 1998.

Mujtaba, M. Shahid: *"Simulation Modelling of a Manufacturing Enterprise with Complex Material, Information and Control Flows"*, Int. Journal of Computer Integrated Manufacturing, Vol. 7, No. 1, pp. 29 - 46, 1994.

Nagel, R: *"Virtual Winners: the Virtual Corporation Could soon Be a Reality"*, International Management, September, pp 22-25, 1993.

Nardi, Bonnie A. and Craig L. Zarmer: *"Beyond Models and Metaphors: Visual Formalisms in User Interface Design"*, IEEE Software Technology Track, Volume 2, 1991.

Naylor, J.B., M.M. Naim and D. Berry: *"Legality: Integrating the Lean and Agile Manufacturing paradigms in the Total Supply Chains"*, International Journal of Production Economics, No. 62 1999.

Neely, A.: *"Measuring Business Performance - Why, What and How"*, Economist Books, London, 1998.

Neely, A., M. Gregory and K. Platts: *"Performance Measurement System Design - A Literature Review and Research Agenda"*, International Journal of Production Management, June 1994.

Neely, Andy and Mike Kennerley: *"Measuring performance in a changing business environment"*, International Journal of Operations and Production Management, Volume 23, Number 2, pp. 213-229, 2003.

Newland Paul, James A. Powell and Chris Creed: *"Understanding Architectural Designers Selective Information Handling"*, Design Studies, Vol. 8, No. 1, January 1987.

Nichols, K.: *"Profit By Design"*, Total Quality Management, pp 105-110, April 1991.

Nielsen, Jakob: *"First rule of usability: Don't listen to users"*, 2001. http://www.useit.com/alertbox/20010805.html

Nixon, Paddy, S. Dobson, V. Wade, J. Fuller, S. Terzis and T. Walsh: *"The Virtues Architectures"*, Department of Computer Science, Trinity College, Dublin, 1998.

Nonaka, I. and N. Konno: *"The Concept of "Ba": Building a Foundation for Knowledge Creation"*. California Management Review, 40(3), pp. 40-54, 1998.

Nonaka, I. and H. Takeuchi: *"The Knowledge Creating Company: How Japanese Companies Create the Dynasties of Innovation"*, New York, Oxford University Press, 1995.

Nutek Inc.: *"Design of Experiments and the Taguchi Approach"*, 1996. http://www.wwnet.com/~rkroy/wp-doe.html#rev.

O'Conner, T.: "Using *Learning Style to Adapt Technology for Higher Education*", 1998. http://web.indstate.edu/ctl/styles/learning.html.

O'Sullivan, David: *"Manufacturing Systems Re-Design: Creating the Integrated Manufacturing Environment"*, PTR Prentice Hall, Englewood Cliffs, 1994.

Ohmae, Kenichi: "The Mind of the Strategist: The Art of Japanese Business", Mc-Graw Hill, 1982.

Oldham, G. R. and A. Cummings: *"Employee Creativity : Personal and Contextual Factors at work"*, Academy of Management Journal, Volume 39, Number 3, pp. 607-634, 1996.

O'Leary, Dennis S.: Presidents Column, *Joint Commission Perspectives*, pp. 2-3, January/February 1996.

Ornstein, Robert: *"The Psychology of Consciousness"*, San Francisco, W.H. Freeman, 1975.

Pannesi, R.T.: *"The Manufacturing Strategy Review Process"*, APICS Conference Proceedings, 1990.

Pawar, K., U. Menon and U., Reetz: *"Human Centred Issues in Concurrent Enterprising"*, in: Weber, F. and K.S. Pawar (Eds.), Proceedings of the 6th. International Conference on Concurrent Enterprising, Toulouse, France, 28 - 30 June 2000.

PBMSIG (Performance Based Management Special Interest Group): *"Guidelines For Performance Measurement"*, 1996. http://nssc.llnl.gov/PBMSig/Review/PMG/DRAFT5BN_3c.html.

Pech Richard J. and Geoffrey Durden: *"Manoeuvre warfare: a new military paradigm for business decision making"*, Management Decision, Volume 41, Number 2, pp. 168-179, 2003.

Peddie, John: *"High Resolution Graphic Display Systems"*, Windcrest/Mc Graw Hill, 1994.

Penrose, Roger: *"The Emperors New Mind: Concerning Computers, Minds and the Laws of Physics"*, Oxford, Oxford Uni. Press, 1989.

Perkins, D.N.: *"The Minds Best Work"*, Cambridge MA, Harvard Uni. Press, 1981.

Pfeifer, T., W. Eversheim, W. Konig, and M. Weck: *"Manufacturing Excellence: The Competitive Edge"*, Chapman and Hall, London, 1994.

Platts, K.W. and M.J. Gregory: *"A Manufacturing Audit Approach to Strategy Formulation,* from *Manufacturing Strategy - Process and Content"*, edited by Christopher A. Voss, Chapman and Hall, London, 1992.

Polanyi, M.: *"The Tacit Dimension"*, Doubleday & Co., 1966.

Porter, Michael E.: *"How Competitive Forces Shape Strategy"*, Marketing Strategy", edited by Dale Littler and Dominic Wilson, Butterworth-Heinemann Ltd., 1995.

Potomac Knowledge Project: *"The Power and Potential of the Knowledge Revolution"*, 1997.
http://knowledgeway.org/milestones/archives/brochure_split/bro_power.htm.

Potter, Richard E. and James L. Ritchie-Dunham: *"CBLE"*, 1995.
http://www.itam.mx/~jrdunham/cble.html

Pratali, Paolo: *"Strategic management of technological innovations in the small to medium enterprise"*, European Journal of Innovation Management, Volume 6, Number 1, pp. 18-31, 2003.

Pratchett, Terry: *"Small Gods"*, Corgi Books, Cox & Wyman Ltd., Reading, 1992.

Pugh, S.: *"Total Design"*, Addison-Wesley, London, 1991.

Pugh, Stuart: *"Total Design – Integrated Methods for Successful Product Engineering"*, Addison-Wesley Publishing, 1990.

Puttick, J.: *"Manufacturing Pull - Manufacturing Push, Springboard for Competitive Advantage"*, PA Consulting Group, London, 1986.

Quinn, J.B. and F.G. Hilmer: *"Strategic Outsourcing"*, Sloan Management Review, Summer, pp 43-55, 1994.

Randall, Neil: *"Making Software easier through Usability Testing"*, PC magazine, 10/6/1998.

Randall, Robert M.: *"The Reengineer"*, Planning Review, May/June 1993.

Ranky, P.G.: *"Concurrent Engineering and Enterprise Modelling"*, Assembly Automation Vol. 14, No 3, pp. 14-21, 1994.

Raskin, Jef: *"Looking for a Humane Interface: Will Computers ever become Easy to Use?"*, Communications of the AMC, Volume 40, Number 2, Feb. 1997.

Raynor, Darrel A. and Richard C. Cohen: *"The Hidden Benefits of Reengineering an Existing Application"*, American Programmer, November 1993.

Reading, Clive: *"Strategic Business Planning: An Action Programme for Forward Thinking Businesses"*, Nichols Publishing, 1993.

Reddy, Sabine and Ram Reddy: *"Competitive agility and the challenge of legacy information systems"*, Industrial Management & Data Systems Volume 102, Number 1, pp. 5-16, 2002.

Rees, Patricia L.: *"User Participation in Expert Systems"*, Industrial Management and Data Systems, Vol. 93, No. 6, 1993.

Reinartz, T., I. Iglezakis, and T. Roth-Berghofer: *"Review and restore for case base maintenance"*, Computational Intelligence: special issue on maintaining CBR systems, 2001.

Richter, R.: *"A critical evaluation of cognitive style assessment"*, Report No. PERS-453, ERIC Document Reproduction Service, 1992.

Reily, R.-F. and R.P. Schweihs: *"Valuing Intangible Assets"*, McGraw-Hill Professional Publishing; New York, 1998.

Rhodes, Jerry: *"Conceptual Tool-making Expert Systems of the Mind"*, Basil Blackwell Ltd., 1991.

Robertson, James: *"Do we need ECO-TAXES?"*, The Word Magazine, Volume 46, No. 9, September 1997.

Rollins, Timothy J.: *"Agents Learning Preferences"*, 1993.
http://wissago.uwex.edu/test/joe/1993summer/rb1.html

Rose, M.: *"Industrial behaviour"*, Penguin, Harmondsworth, 1988.

Rosenstein, Aviva: *"Do You Know Who Your Users Are?"*, 2001.
http://www.classicsys.com/classic_site/articles/article_4-01.html

Ross, B.H.: *"Case-Based Reasoning Workshop"*, DARPA, pp. 144-147, 1989.

Roth, Aleda V., Craig A. Giffi and Gregory M. Seal: *"Operating Strategies for the 1990s: Elements Comprising World-Class Manufacturing"*, Manufacturing Strategy - Process and Content, edited by Christopher A. Voss, Chapman and Hall, London, 1992.

Rothschild, Michael: *"Cro-Magnons Secret Weapon"*, Forbes ASAP, Sept. 1993.
http://www.bionomics.org/text/resource/articles/ar_020.htm.

Roy, Marie Christine, Oliver Dewit and Benoit A. Aubert: *"The impact of web usability on trust in web retailers"*, Internet Research: Electronic Networking Applications and Policy, Volume 11 Number 5 2001.

Saaty, T.L.: *"How to make a decision, The Analytic Hierarchy Process"*, European Journal of Operational Research, 48, 1990, pp. 9-26.

Saaty, T.L.: *"The Analytic Hierarchy process"*, McGraw-Hill, New York, 1980.

Sackett, P.J.: *"Manufacturing Strategy Decision Support Tools - Linking Corporate Vision to Measures of Manufacturing Performance in Concurrent Engineering"*, Proceedings 1996 ASST, Concurrent Product and Process Design and Development, Damascus, pp. 47 - 67, 1996.

Sackett, P.J.: *"World Class Manufacturing - Linking Future Scenarios, Strategies and Technologies"*, Proceedings of the PRO E UC, Johannesburg, SA, August, 1996.

Sackett, P.J.: *"Manufacturing Strategy Theories, Models and Frameworks"*, AMBITE document, AMBITE/CU/WP1/2/05/D/F/1.0/F, 1994.

Salama Alzira and Mark Easterby-Smith: *"Cultural Changes and Managerial Careers"*, 1995.
http://www.mcb.co.uk/services/articles/liblink/pr/salama.html.

Salomon, Gitta: *"New Uses for Colour"*, The Art of Human-Computer Interface Design, Addison-Wesley Publishing Company, 1990.

Sanderson, S.M. and G.A. Luffman: *"Strategic Planning and Environmental Analysis"*, Marketing Strategy, edited by Dale Littler and Dominic Wilson, Butterworth-Heinemann Ltd., 1995.

Schactman, Eric: *"Model"*, 1996.
http://capita.wustl.edu/ME567_Informatics/concepts/model

Scherr, A. L.: *"A New Approach to Business Processes"*, IBM Systems Journal, Volume 32, No. 1, 1993.

Schmenner, R.: *"Production/Operations Management"*, Macmillan, New York, 1990.

Schonberger, R.J.: *"World Class Manufacturing Casebook - Implementing JIT and TQC"*, The Free Press, 1987.

Schrage, D.P.: *"Concurrent Design: A Case Study, Concurrent Engineering"*, Automation, Tools and Techniques, A. Kusiak (editor), John Wiley ands Sons 1993.

Schutzer, Daniel: *"Business Decisions with Computers: New Trends in Technology"*, Van Nostrand Reinhold, New York, 1991.

Sculley, John: *Odyssey: "From Pepsi to Apple"*, New York, Harper & Row, 1992.

SECRC: *"Concurrent Engineering"*, 1996. http://www.ecrc.edu/ce.

Seliger, G., S. Krüger, and Y. Wang: *"Integrated Information Modelling for Simultaneous Assembly Planning"*, Institute for Machine Tools and Manufacturing Technology, Berlin, 1992 (not published).

Selker T. and W. Burleson: *"Creativity and Interface: Introduction"*, Communications of the AMC, Volume 45, Number 10, 2002.

Senge, Peter M.: *"The Fifth Discipline: The Art and Practice of the Learning Organisation"*, Century Business, 1992.

Senn, James A: *"Information Systems in Management"*, Wadsworth Publishing Co., 1990.

Shapiro, Michael A., and Daniel G. Mc Donald: *"I'm Not a Real Doctor, but I Play One in Virtual Reality: Implications of Virtual Reality for Judgements about Reality"*, Journal of Communication, Vol. 42, Part 4, 1992.

Shaw, N. C: *"Knowledge Management Basics"*, ICASIT- International Center for Applied Studies in Information Technology, 2003.

Shenas, D.G. and S. Derakhshan: *"Organisational Approaches to the Implementation of Simultaneous Engineering"*, International Journal of Operations and Production Management, Vol. 14, No. 10, pp 30-43, 1994.

Shneiderman, B.: *"Creativity Support Tools"*, Communications of the AMC, Volume 45, Number 10, 2002.

Simon, H.A.: *"Making Management Decisions: The Role of Intuition and Emotion"*, Academy of Management Executive, pp. 57 - 64, February 1987.

Skinner, W.: *"Manufacturing in the Corporate Strategy"*, Wiley 1978.

Skinner, Wickham: *"Manufacturing-Missing Link in Corporate Strategy"*, Harvard Business Review, pp. 136 – 145, May-June, 1969.

Skyrme, D.: *"Networking to a Better Future – Management Insights"*, 1996. http://www.hiway.co.uk/skyrme/insights/insights.html.

Slessk, David: *"Learning and Visual Communication"*, Halsted Press, 1981.

Smith Ring, P. and Andrew H. Van de Ven: *"Structuring Cooperative Relationships Between Organisations"*, Strategic Management Journal; Vol. 13, pp 483-498, 1992.

Smith, Sidney L., and Jane M. Mosier: *"Guidelines for Designing User Interface Software"*, August 1986. ftp://ftp.cis.ohio-state.edu/pub/hci/Guidelines/guidelines

So, Stella and Malcolm Smith: *"The impact of presentation format and individual differences on the communication of information for management decision making"*, Managerial Auditing Journal, Volume 18, Number 1, pp. 59-67, 2003.

Sohlenius, G.: *"Concurrent Engineering"*, Annals of the CIRP, Vol. 41, 1992.

Spitz, S. Leonard: *"Machine Vision for Inspection of Hybrid Circuits earns mixed reviews"*, Electronic Packaging and Production, pp. 56 - 60, October 1990.

SRC: "The History of the SRC[15]", 1997. http://www.src.org/about/history.cgi/32a2606c8b962814.

Srinivas, H.: *"Knowledge Management"*, The Global Research Center, 2003.

Stacy, Webb: *"Cognition and Software Development"*, Communications of the ACM: Cognition and Software Development, Volume 38, No. 6, June 1995.

Stalk, G. Jr. and A.M. Webber: *"Japan's Dark Side of Time"*, Harvard Business Review, July-August, pp 93-102, 1993.

Stanek, mary Beth: *"The need for global managers: A business necessity"*, Management Decision, Volume 38, Number 4, pp. 232 - 242, 2000.

Steele, Lowell W.: *"Managing Technology - The Strategic View"*, McGraw-Hill, New York, 1989.

Stein, Randall: *"Rattling the Cage: Preparing Information Systems Personnel for Business Process Reengineering"*, American Programmer, November 1993.

Stewart, Tom: *"An Employers Guide to the Display Screen Regulations"*, 1995. http://www.system-concepts.com/stds/hse5.Html#RTFToC61

Stoddard, Donna B. and Sirkka L. Jarvenpaa: *"Business Process Redesign: Tactics for Managing Radical Change"*, Journal of Management Information Systems, Volume 12, No. 1, pp. 81 - 107, Summer 1995.

Stonehouse George and Jonathon Pemberton: *"Strategic planning in SMEs - some empirical finding"*, Management Decision, Volume 40, Number 9, pp 853-861, 2002.

Stoney, Christopher: *"Strategic management or strategic Taylorism?"*, The International Journal of Public Sector Management, Volume 14, Number 1, pp. 27-42, 2001.

Strong, G.W.: *"New Directions in Human-Computer Interaction"*, Education, Research and Practice, 1994. http://www.sei.cmu.edu/arpa/hci/directions

Sure, Y., S. Staab and R. Studer: *"Methodology for Development and Employment of Ontology based Knowledge Management Applications"*, Special Section on Semantic Web and Data Management, SIGMOD Record(4), 2002.

Szirbik, N.B. and H.S. Jagdev: *"The Future of IT Systems for Virtual Enterprises: Product-Oriented Agent Providers"*, 8th IEEE International Conference on Emerging Technologies and Automation (ETFA2001), Antibes Juan-les-pins, France, 15-18 October 2001.

Tamimi, Nabil and Rose Sebastianelli: *"How product quality dimensions relate to defining quality"*, International Journal of Quality & Reliability Management, Volume 19, Number 4, pp. 442-453, 2002.

Taylor, F.W.: *"Principles of Scientific Management"*, New York, Harper, 1911.

Thacker, Rebecca A.: *"Team leader style: enhancing the creativity of employees in teams"*, Training for Quality, Volume 5, Number 4, pp. 146-149, 1997.

The Economist Intelligence Unit: "Executive Information Systems: Management Guide", July 1991.

Thoben, K-D, F. Weber and G. Giarda: *"Accelerating the Exchange of Information and Experience about Concurrent Engineering"*, in: Martensson, N; R.

---

[15] SRC stands for the Semi-conductor Research Co-operative.

Mackay and S. Björgvinsson (Eds.), Changing the Ways We Work; Advance in Design and Manufacturing (Vol. 8); 1998.

Thoben, K-D and F. Weber: *"Designing Information and Communication Structures for Concurrent Engineering - Findings from the Application of a Formal Method"*, Proceedings of International Conference on Engineering Design, ICED99, Munich, August 24 26, 1999.

Thoben, K.-D., Jagdev, H. and Eschenbaecher, J.: *"Extended Products: Evolving Traditional Product concepts"*, Proceedings of the 7th International Conference on Concurrent Enterprising: Engineering the Knowledge Economy through Co-operation. Bremen Germany 27-29th June 2001.

Thoben, K.-D., Jagdev, H.S. and J. Eschenbächer: *"Using e-business to provide Extended Products*, Automation 2001, Helsinki, Finland, 5-6 September 2001.

Thoben, K.-D., J. Eschenbächer and H.S. Jagdev: *Emerging Concepts in E-Business and Extended Products"*, In E-Business Applications, J Gasos and K-D Thoben (Editors), Springer-Verlag, Berlin, 2003.

Thommessen, C.: *"Network computing creating the new era of Extended Enterprise"*, Financial Times, June 5, 1996.

Tipnis, V. A.: *"Evolving Issues in Product Life Cycle Design"*, Annals of the CIRP, Vol. 42, No. 1, pp. 169-173, 1993.

Transformation: *"IBM Systems Journal"*, Volume 32, No. 1, pp. 65 - 79, 1993.

Travica, B. and B. Cronin: *"The Argo: A Strategic Information System for Group Decision Making"*, 1996. http://nickel.ucs.indiana.edu/~btravica/argo.html

Trygg, L.: *"Concurrent Engineering Practices in Selected Swedish Companies: A Movement or an Activity of the Few?"*, Journal of Product Innovation Management, Vol. 10, pp 403-415, 1993.

Tuomi, I.: *"Data Is More Than Knowledge: Implications of the Reversed Knowledge Hierarchy for Knowledge Management and Organisational Memory"* The 32nd Hawaii International Conference on System Sciences, Maui, Hawaii, 1999.

Turino, J.: *"Managing Concurrent Engineering. Buying Time To Market"*, New York, Van Nostrand Reinhold, 1992.

Twiss, B. and M. Goodridge: *"Managing Technology for Competitive Advantage"*, Pitman: London, 1989.

Udo, Goodwin G.: *"Using analytic hierarchy process to analyze the information technology outsourcing decision"*, Industrial Management & Data Systems, Volume 100, Number 9, pp. 421-429, 2000.

USDOE (United States Department of Energy): *"How to Measure Performance: A Handbook of Techniques and Tools"*, 1997.

UTS School of Engineering: *"Information Systems Engineering"*, 1996. http://www.ee.uts.edu.au/eeo/courses/pg/ise.html.

van Assen, M.F., E.W. Hans and S.L van de Velds: *"An agile planning and control framework for customer-order driven discrete parts manufacturing"*, International Journal of Agile Management Systems, Vol.2 No. 1, 2000.

Van Weele, A.J.: *"Purchasing management: Analysis, Planning and Practice"*, Chapman and Hall. London, UK, 1994.

Vesey, J.T.: *"Time-to-Market: Put Speed in Product Development"*, Industrial Marketing Management, Vol. 21, pp. 151-158, 1992.

Vickers, Goeffrey: *"The Vickers Papers"*, Open Systems Group, 1984.

von Luck, K. B. Nebel, C. Peltason and A. Schmiedel: *"The Anatomy of the BACK system"*, KIT Rept. 41, Fachbereich Informatik, Technische University Berlin, Berlin (1987).

Waggoner, D.B., A. Neely and M. Kennerley: *"The forces that shape organisational performance measurement systems: An interdisciplinary review"*, International journal of production economics, 60, pp. 53-60, 1999.

Wang, Charles B., CA-KBM: *"Knowledge isn't only power, its Profits"*, 1997. http://www.cai.com/products/kbm.htm.

Watson, Hugh J., Joyce Elam, Jenny Harris, Ellen Hertz, R. Kelly Rainer, Ronald S. Swift and Douglas R. Vogel (Panel): *"A Research Agenda for Executive Information Systems"*, Proceedings of the Twenty-Sixth Hawaii Annual International Conference on System Sciences, Volume 3, 1993.

Webb, Ian: *"Thinking Strategy"*, Industrial Society Press, 1989.

Weber, F. and Pawar, K.S. (Eds.): *"Enhancing Business Competitiveness through Sharing Experiences between Research and Industry"*, Proceedings of the 6th. International Conference on Concurrent Enterprising, Toulouse, France, 28 - 30 June 2000.

Weber F., B Brederhorst and R. J. Barson: *"The application of case based reasoning to decision support in new product development"*, Integrated Manufacturing Systems, Volume 14, Number 1, pp. 36-45, 2003.

Weber, S.F.: *"A Modified Analytic Hierarchy Process for Automated Manufacturing Decisions"*, Interfaces, 23, 4, pp 75-84, July-August 1993.

Werner, P.: *"Car Industry"*, Human Systems Management, Vol. 12, No. 1, pp. 65 - 73, 1993.

Wheelwright, Steven C. and Robert H. Hayes: *"Competing through Manufacturing"*, Harvard Business Review, January - February 1985.

Wheelwright, Steven C.: *"Strategy, Management and Strategic Planning Approaches"*, Interfaces, Vol. 14, No. 1, pp. 19 - 33, 1984.

Wilkinson, Richard: *"Reengineering Industrial Engineering in Action"*, Industrial Engineering, August 1991.

Williamson, O.E.: *"Markets and Hierarchies"*, The Free Press, New York, NY, USA, 1975.

Winner, R.I., J.P. Pennell, H. E. Bertrnad and M.M.G. Slusarczuk: *"The Role of Concurrent Engineering in Weapons System Acquisition"*, (IDA Report R-338), Alexandria, VI, Institute for Defense Analysis, 1988.

Winograd, Terry: *"Evolving a Software Design Curriculum"*, 1994. http://www-pcd.stanford.edu/asd/info/articles/design-curriculum.html.

Winograd, Terry: *"From Programming Environments to Environments for Designing"*, Communications of the ACM: Cognition and Software Development, Volume 38, No. 6, June 1995.

Womack, J.P, D.T. Jones and D. Ross: *"The Machine that Changed the World"*, Harper Perennial, USA, 1991.

Womack, J.P and D.T. Jones: *"Lean Thinking"*, Simon & Schuster, 1996.

Woo, Andrew, Pierre Poulin and Alain Fournier: *"A Survey of Shadow Algorithms"*, IEEE Computer Graphics and Applications, Nov. 1990.

Woodstock, M. and Francis, D.: *"The Unblocked Manager; A practical guide to self-development"*, Gower, Hampshire, 1993.

Wooley, Benjamin: *"Virtual Worlds"*, Penguin Books, 1993.

Xiao, Y., and C.F. Mackenzie: *"Decision Making in Dynamic Environments: Fixation Errors and their Causes"*, 1995.
http://audio.ab.umd.edu/fixation/fixation.html.

Yeh, *Chung-Hsing: "A customer-focused planning approach to make-to-order production"*, Industrial Management & Data Systems, Volume 100, Number 4, pp. 180-187, 2000.

Yu, B., J.A. Harding and K. Poppelwell: *"Supporting enterprise design through multiple views"*, International Journal of Agile Management Systems, Vol. 2, No. 1, 2000.

Zairi, M.: *"Measuring performance for business results"*, Chapman and Hall, London, 1994.

Zhao, J., W. M. Cheung and R.I.M. Young: *"A consistent manufacturing data model to support virtual enterprises"*, International Journal of Agile Management Systems, Volume 1, Number 3, pp. 150-158, 1999.

Appendix

List of Performance Measures

This appendix gives a list of performance measures for the various macro business processes identified in chapters 3 and 8. This list of performance measures is not exhaustive but is intended as a reference list of measures that can be used to measure specific business processes along the five competitive dimensions identified earlier, namely: time, cost, quality, flexibility and the environment.

Customer Order Fulfilment Process - Time
- Average customer order fulfilment time
- Average customer order delivery time
- Average order processing time
- Average order confirmation time
- Average time taken to develop a master production schedule (MPS)
- Average time taken to develop a forecast
- Average customer query response time
- Average time taken to generate a rough cut capacity plan (RCCP)
- Average time taken to maintain the MPS
- Average time taken to run the MRP
- Average time taken to generate a capacity plan
- Average time taken to generate a factory wide schedule
- Average time taken to receive a customer order
- Average time taken to package a customer order
- Average time taken to ship a customer order
- Average time taken to configure a customer order
- Average time taken to plan the assembly of an order
- Average time taken to plan a customer project

Customer Order Fulfilment Process - Cost:
- Cost of products sold per unit time
- Cost of warranty per unit time
- Cost of warranty as a percentage of cost of sales
- Cost of outstanding orders per unit time
- Cost of damaged orders per unit time
- Cost of material inventory
- Average cost per purchase order

- Average cost of developing a master production schedule (MPS)
- Average cost of developing a forecast
- Average cost of carrying out a rough cut capacity plan (RCCP)
- Average cost of maintaining a MPS
- Average cost of generating a capacity plan
- Average cost of changing a planned order
- Average cost of generating a factory wide schedule
- Average cost of receiving a customer order
- Average cost of delivering a customer order
- Average cost of packaging a customer's order
- Average product configuration cost
- Average cost of planning a customer project

Customer Order Fulfilment Process - Quality:
- Number of customer complaints per unit time
- Number of warranty period extensions per unit time
- Number of warranty claims per unit time
- Number of shipments per unit time
- Number of customer returns per unit time
- Number of outstanding orders per unit time
- Number of damaged orders per unit time
- Number of lost orders per unit time
- Number of customers connected with EDI
- Number of order changes per customer
- Number of customer surveys
- Number of partnerships formed with customers per unit time
- Number of hours spent under staff exchange
- Number of hours spent in training per employee per unit time
- Number of employees in project teams
- Education/Training budget as a percentage of cost of sales
- Number of suggestions per employee
- Average number of levels in BOM
- Average deviation between forecast and actual sales
- Percentage of orders delivered incorrectly
- Percentage of MPS generated correctly
- Number of updates required per MPS
- Number of changes made to the BOM per MRP run
- Percentage of cell schedules generated correctly
- Percentage of customer orders received correctly
- Percentage of complete orders delivered
- Percentage of orders delivered on schedule
- Percentage of orders packaged correctly
- Percentage of incompatible product options selected by the customer
- Number of changes made to a customer project after it has been planned

Customer Order Fulfilment Process - Flexibility:
- Number of changes to schedule per unit time

- Number of inventory turns per unit time
- Ration of purchase orders placed to the purchasing department head count
- Ratio of purchase order errors to the number of purchase orders audited
- Percentage of people in teams
- Number of hours spent in training per employee per unit time
- Number of job classifications
- Number of employees in project teams
- Education/training budgets as a percentage of cost of sales
- Number of orders delivered
- Time horizon of MPS
- Number of MPS generated per quarter
- Percentage of MPS updates requested that are carried out
- Number of changes made to the BOM per MRP run
- Percentage increase/decrease in orders waiting to be received
- Number of order awaiting delivery
- Percentage increase/decrease in product orders waiting to be configured
- Number of customer projects planned per unit time

Customer Order Fulfilment Process - Environment:
- Volume of packaging used to ship products
- Percentage of regulations governing this area that have been met
- Volume of paper produced
- Volume of packaging used per delivery
- Level of CO transmissions per delivery
- Percentage of products re-cycled at end of life

Vendor Supply Process - Time:
- Average materials delivery lead time
- Average vendor query response time
- Average vendor certification time
- Deviation between vendor's material and the industry norm
- Number of days material is out of stock
- Vendor approval time
- Vendor selection time
- Vendor contract development time
- Material ordering time
- Purchase order generation time
- Time taken to receive arriving materials
- Arriving materials inspection time

Vendor Supply Process - Cost:
- Cost of material inventory
- Cost of vendor non-conformance
- Cost of out of stock per unit time
- Cost of materials received from vendors
- Cost of materials received from approved vendors
- Cost of materials ordered from vendors

- Vendor selection cost
- Vendor approval cost
- Vendor contract development cost
- Purchase order generation cost
- Arriving materials inspection cost
- Cost of receiving arriving materials

Vendor Supply Process - Quality:
- Number of partnerships formed with vendors per unit time
- Percentage of material orders that are overdue
- Total number of vendor deliveries received
- Percentage of on-time vendor deliveries
- Percentage of material orders received late
- Percentage of material orders received that are damaged
- Percentage of vendor orders that are lost
- Percentage of vendors connected via electronic data interchange (EDI)
- Total number of vendors
- Percentage of vendors that are approved
- Percentage of vendors that are local
- Percentage of suppliers in project teams
- Number of new product improvement proposals per supplier
- Number of supplier partnerships per unit time
- Total number of shipments received per unit time
- Number of different material part numbers
- Average number of different parts supplied by each vendor
- Number of hours spent under staff exchange
- Number of hours spent in training per employee per unit time
- Percentage of employees in project teams
- Education/training budgets as a percentage of cost of sales
- Number of suggestions per employee
- Percentage deviation between the vendor and market prices
- Percentage of arriving materials that pass quality inspection
- Percentage of cost saving initiatives suggested by the vendor
- Number of complaints made against vendors
- Percentage of errors made in the vendors booking-in procedures
- Percentage deviation of vendors from agreed delivery schedule
- Number of different part numbers supplied per vendor
- Number of approved vendors
- Number of vendors supplying the ordered materials
- Percentage of approved vendors that have quality problems
- Percentage of approved vendors that are dropped
- Percentage of potential vendors that are chosen
- Percentage of material improvements suggested per vendor
- Percentage of vendors who lost "Approved" status
- Percentage of vendor contracts developed correctly
- Percentage of POs sent to the correct vendors
- Percentage of materials correctly ordered from vendors

- Percentage of material correctly identified by enterprise
- Percentage of material orders supplied correctly
- Percentage of materials arriving damaged
- Percentage of materials damaged during unpacking

Vendor Supply Process - Flexibility:
- Ratio of purchase orders placed to the purchasing department head count
- Ratio of purchase order errors to the number of purchase orders audited
- Number of partnerships formed with vendors per unit time
- Average number of vendors per product
- Number of hours spent in training per employee per unit time
- Number of job classifications
- Number of employees in project teams
- Education/training budgets as a percentage of cost of sales
- Number of material orders received per unit time
- Number of different part number supplied per vendor
- Percentage of vendor contracts developed per unit time
- Number of POs generated per unit time
- Number of POs sent per unit time
- Number of material deliveries outstanding per unit time
- Percentage increase/decrease in number of deliveries waiting to be received
- Percentage increase/decrease in material orders waiting to be inspected

Environment:
- Percentage of regulations governing this area that have been met
- Volume of paper produced per unit time
- Percentage of vendors supplying re-cycleable material
- Percentage of vendors that re-cycle their material
- Volume of packaging left for disposal
- Percentage of environmentally unfriendly components ordered from vendors

Manufacturing Process - Time:
- Total production lead time
- Production process time
- Average set-up time per product
- Time lost due to bottlenecks
- Number of overtime hours per unit time
- Percentage machine idle time
- Total unplanned machine down time per unit time
- Manufacturing lead time
- Process down time due to stock delays
- Mean time between failures
- Mean time to repair
- Number of hours spent in training per employee per unit time
- Time lost due to staff conflicts
- New process/equipment introduction time
- Total annual maintenance time

Manufacturing Process - Cost:
- Cost of defective product due to process failure
- Cost of quality function
- Scrap cost per unit time
- Rework cost per unit time
- Cost of failure of product design
- Cost of inventory
- Cost of work in progress inventory
- Cost of equipment used in manufacturing process
- Appraisal cost
- Manufacturing labour cost
- Manufacturing overhead cost
- Average rework cost per customer return
- Capital investment cost as a percentage of total cost
- Manufacturing material cost
- Cost of maintenance
- Maintenance cost per unit output price
- Energy consumption cost per unit time
- Cost of manufacturing the product
- Cost of resources used to produce product
- Warranty cost attributed to manufacturing

Manufacturing Process - Quality:
- Number of products produced
- Total number of stoppages
- Number of product defects per unit time
- Direct labour productivity
- Indirect labour productivity
- Total number of stoppages per machine
- Number of measurable processes
- First time yield
- Capacity utilisation
- Percentage size of manufacturing shop floor
- Percentage size of repair area
- Percentage of employees in project teams
- Number of employees per supervisor
- Number of injuries per unit time
- Number of days lost due to absenteeism
- Percentage of manufactured product that pass quality tests
- Percentage of manufacturing problems that are process related
- Percentage of set-ups performed correctly

Manufacturing Process - Flexibility:
- Percentage production output to planned output
- Number of suggestions per employee
- Number of products manufactured
- Number of different products manufactured

- Number of products manufactured from different product families
- Number of set-ups required per product family

Manufacturing Process - Environment:
- Level of toxic waste per product
- Level of toxic waste per process
- Percentage of processes that are environmentally unfriendly
- Number of environmentally unfriendly substances used in the manufacturing process
- Percentage of regulations governing this area that have been met
- Percentage of environmentally unfriendly components used per product

Design Co-ordination Process - Time:
- Average development time per new product (months)
- Prototype lead time (months)
- Time to market for a new product
- Average product development time
- New product introduction time
- New model introduction time
- New process/equipment introduction time
- Average lead time to delivery of co-engineered product by supplier
- Life of product commercially
- Life of product design

Manufacturing Process - Cost:
- New process development cost
- Average development cost per new product
- Cost per drawing
- Product development cost
- Product R&D costs

Manufacturing Process - Quality:
- Percentage of new components in the new product
- Average number of levels in BOM
- Average number of engineering hours per new product (thousands)
- Ratio of delayed products to new products introduced
- Number of new products introduced per unit time
- Number of product upgrades introduced per unit time
- Percentage of CAD hours used
- Level of bought in parts
- Average number of components per product
- Average number of product options available
- Number of components with shared design per product
- Number of drawings produced per product
- Number of design defects identified per product
- Number of engineering change orders received per product
- Number of different parts used per product

- Number of modules per product
- Percentage of standard parts in new designs
- Percentage of supplier propriety parts in new designs
- Number of engineering changes
- Percentage of shared parts
- Supplier share of engineering
- Percentage of product design carried out by suppliers
- Percentage of supplier designed parts in new product
- Percentage of suppliers in project team
- Percentage of resources allocated to R&D
- Number of hours spent under staff exchange
- Percentage of employees in project teams
- Number of hours spent in training per employee per unit time
- Education/training budget as a percentage of cost of sales
- Productivity of design loop employees
- Number of job classifications

Manufacturing Process - Flexibility:
- Number of design changes per product
- Number of new products introduced annually
- Number of different products made

Manufacturing Process - Environment:
- Percentage of product design that is re-cycleable
- Percentage of products re-cycled at end of life
- Percentage of products containing toxic material
- Percentage of vendors supplying re-cycleable material
- Percentage of reused components per product
- Percentage volume of material re-usable
- Percentage volume of material re-cycleable
- Level of toxic waste per product
- Level of toxic waste per process

Co-Engineering Process - Time:
- Average lead time to delivery of co-engineered product by supplier
- Number of hours spent in training per employee per unit time
- Time taken to transmit design information to co-engineering vendor
- Time taken to receive co-engineering design information from vendor(s)

Co-Engineering Process - Cost:
Not applicable as the majority of the cost related performance measures in this process will be borne by the vendor(s) performing the co-engineering. Their costs will probably not be relevant to the enterprise that has contracted this work. Also, the co-engineering process is the design process for the vendor and the performance measures used by the vendor may be quite similar to those used in the enterprises design process.

Co-Engineering Process - Quality:
- Percentage of new design ideas suggested by vendors
- Percentage deviation of vendors from agreed co-engineering delivery schedule
- Percentage of co-engineering vendors linked using EDI
- Percentage of shared parts
- Number of components with shared design per product
- Percentage of supplier propriety parts in new designs
- Percentage of engineering carried out by suppliers
- Percentage of supplier designed parts in new product

Co-Engineering Process - Flexibility:
- Rate the supplier's assistance in solving technical problems
- Number of hours spent under staff exchange
- Percentage of people in teams
- Number of suppliers in project team

Co-Engineering Process - Environment:
- Percentage of co-engineered design that is re-cycleable
- Percentage of reused components per co-engineered design
- Percentage of supplier specific parts in co-engineered design

Glossary of Acronyms

ACE	Aggregated Causal Effect
ADOC	Aggregated Degree Of Confidence
AEDC	Australian Electronics Development Centre
AHP	Analytic Hierarchy Process
AMBIT	Advanced Manufacturing Business Implementation Tool
ATO	Assemble To Order
B2B	Business to Business electronic commerce
B2C	Business to Consumer electronic commerce
BDF	Business Design Facility
BPR	Business Process Re-engineering
C	Cost
CAD	Computer Aided Design
CAM	Computer Aided Manufacturing
CBLE	Computer Based Learning Environment
CBR	Case Based Reasoning
CCOQ	Quality of collected customer order
CCOT	Time to Collect Customer Order
CE	Concurrent Engineering
CE function	Causal Effect function
CF	Confidence Function
CIM	Computer Integrated Manufacturing
CIM-OSA	Computer Integrated Manufacturing – Open Systems Architecture
COB	Chip On Board
COF	Customer Order Fulfilment
COFT	Time to Fulfil Customer Order
CRs	Customer Requirements
CSF	Critical Success Factor
CSFs	Critical Success Factors
DCOT	Time to Deliver Customer Order

DEC	Digital Equipment Corporation
DFA	Design For Assembly
DFE	Design For the Environment
DFM	Design For Manufacture
DIP	Dual Inline Package
DOC	Degree Of Confidence
DOE	Design Of Experiments
DSS	Decision Support System
E	Environment
EIS	Executive Information System, Enterprise Information System
EMS	Enterprise Modelling System
EOQ	Quality of Entered Order
EOT	Time to Enter Order
ERP	Enterprise Resource Planning
ETO	Engineer To Order
F	Flexibility
FMEA	Failure Modes Effect Analysis
FMS	Flexible Manufacturing Systems
FPT	Fine Pitch Technology
GUI	Graphical User Interface
HCI	Human Computer Interface
I/O	Input/Output
IC	Integrated Circuit
ICE	Independent Causal Effect
ICT	Information and Communications Technologies
IDEA	Interface Design Aid
IDEA tool	Interface Design Aid modelling tool
ILB	Inner Lead Bonding
IOQ	Quality of Intake Order
IOT	Time to Intake Order
IT	Information Technology
KASPER	Knowledge Based Advisor and Strategic Programme Evaluator
KPV	Key Programme Variable
KT/PVs	Key Technology and Key Programme Variables
KTV	Key Technology Variable
LEAGILE	Lean and Agile
LO	Linking Object
LP	Lean Production

LSI	Large Scale Integration
M_MOMPs	Manufacturing model MOMPs
MCM	Multi Chip Modules
MCM-C	Multi Chip Modules – Ceramic dielectric
MCM-D	Multi Chip Modules – Deposited dielectric
MCM-L	Multi Chip Modules – Laminated dielectric
MCOT	Time to Manufacture Customer Order
MOMP	Measure of Manufacturing Performance
MP	Manufacturing Process chain
MRP	Materials Requirements Planning
MRPII	Manufacturing Resource Planning
MTO	Make To Order
MTS	Make To Stock
OMM	Organisation Modelling Method
PCB	Printed Circuit Board
PCOT	Time to Process Customer Order
PDM	Product Data Management
PM	Performance Measurement
PO	Purchase Order
POT	Time to Process Order
PROTIMA	Programme/Technology Implementation Analyser
PTH	Plated Through Hole
PWB	Printed Wiring Board
Q	Quality
QFD	Quality Function Deployment
ROQ	Quality of Received Order
ROT	Time to Receive Order
SCOQ	Quality of Ship Customer Order
SCOT	Time to Ship Customer Order
SE	Sequential Engineering
SMART	Specific, Measurable, Attainable, Realistic and Timed
SMT	Surface Mount technology
SMT-SC	Surface Mount Technology – Single Chip
SPC	Statistical Process Control
SPI	Strategic Performance Indicator
SSI	Small Scale Integration
SWOT	Strengths, Weaknesses, Opportunities and Threats analysis
T	Time

T_MOMPs	Technology model MOMPs
TAB	Tape Automated Bonding
TQM	Total Quality Management
VLSI	Very Large Scale Integration
VR	Virtual Reality
VSC	Vendor Supply Chain

Index

CD-ROM Disclaimer